Svetlin Georgiev
Editor

Numerical Optimization on Time Scales

DOI: https://doi.org/10.52305/LGNH2335

Library of Congress Cataloging-in-Publication Data

ISBN: 979-8-89530-466-2 (Softcover)
ISBN: 979-8-89530-503-4 (eBook)

Published by Nova Science Publishers, Inc. † New York

Contents

Preface

Time scale theory was first initiated by Stefan Hilger in 1988 in his PhD thesis to unify both approaches of dynamic modelling: difference and differential equations. Similar ideas have been used before and go back in the introduction of the Riemann-Stieltjes integral which unifies sums and integrals. Many results to differential equations carry over easily to corresponding results for difference equations, while other results seem to be totally different in nature. Because of these reasons, the theory of dynamic equations is an active area of research. The time scale calculus can be applied to any fields in which dynamic processes are described by discrete or continuous time models. So, the calculus of time scales has various applications involving non-continuous domains such as certain bug populations, phytoremediation of metals, wound healing, maximization problems in economics and traffic problems.

This book is devoted on solving of some optimization problems on time scales. Because of the wide (and growing) use of optimization in science, engineering, economics, and industry, it is essential for students and practitioners alike to develop an understanding of optimization algorithms on time scales. Knowledge of the capabilities and limitations of these algorithms leads to a better understanding of their impact on various applications, and points the way to future research on improving and extending optimization algorithms and software.

The book contains five chapters. In Chapter 1 some central concepts of practical optimization techniques are considered. In Chapter 2, we discuss some line search methods. The Wolfe conditions, strong Wolfe conditions and Goldstein conditions are introduced. Existence of step lengths that satisfy these conditions is proved. Closedness and convergence of line search methods are established. Rates of line search methods are investigated. The Newton method is introduced and some of its properties are deduced. A quasi Newton method is introduced and it is proved that it is superlinearly convergent. They are described line search procedures based on interpolation of known function and derivative values of the function. In Chapter 3, we introduce some conjugate direction methods. The conjugate direction theorem is proved. Some descent properties of the conjugate direction method are given. In particular, the expanding subspace theorem is deduced. The conjugate direction algorithm is described and a variant of the conjugate direction theorem is formulated and proved. The C-G method is introduced as a generalization of the conjugate direction method. Some bounds of convergence are deducted. Chapter 4 is devoted on the modified Newton method. We derive its convergence rate by employing the analysis of the method of steepest descent. We five algorithms for construction of the inverse of the Δ-Hessian in the case when it is a constant positive definite and symmetric matrix. We propose two procedures: a rank one procedure and a rank two procedure. In Chapter 5, we investigate minimization problems with equality and inequality constraints. They are given some first order and second order necessary and sufficiency conditions when a point is a local minimum point. Dual problems are introduced and the duality theorem is proved.

This book is addressed to a wide audience of specialists such as mathematicians, physicists, engineers and biologists. It can be used as a textbook at the graduate level and as a reference book for several disciplines.

Paris, October 2024 *Svetlin G. Georgiev*

Chapter 1

Introduction

One common application of calculus is calculating the minimum or maximum value of a function. For example, companies often want to minimize production costs or maximize revenue. In manufacturing, it is often desirable to minimize the amount of material used to package a product with a certain volume. Optimization is the minimization or maximization of a function subject to constraints on its variables. If $\mathbb{T}_j$, $j \in \{1,\ldots,n\}$, are time scales with forward jump operators, backward jump operators, delta differentiation operators and nabla differentiation operators σ_j, ρ_j, Δ_j and ∇_j, $j \in \{1,\ldots,n\}$, respectively, $\sigma = (\sigma_1,\ldots,\sigma_n)$, $\rho = (\rho_1,\ldots,\rho_n)$, $\Delta = (\Delta_1,\ldots,\Delta_n)$, $\nabla = (\nabla_1,\ldots,\nabla_n)$, $\Lambda^n = \mathbb{T}_1 \times \ldots \times \mathbb{T}_n$, $x = (x_1,\ldots,x_n) \in \Lambda^n$ is the vector of independent variables, f is a scalar function of x that we want to maximize or minimize, called the objective function, and c_j are constraint functions, which are scalar functions of x, that define certain equations and inequalities that the unknown vector x must satisfy. Using this notation, the optimization problem can be written as follows

$$\min_{x \in \Lambda^n} f(x) \quad \text{subject to} \quad \begin{aligned} c_j(x) &= 0, \quad j \in I_1, \\ c_j(x) &\geq 0, \quad j \in I_2. \end{aligned} \tag{1.1}$$

Here, I_1 and I_2 are sets of indices for equality and inequality constraints, respectively.

Example 1 *Let $\mathbb{T}_1 = 2^{\mathbb{N}_0}$, $\mathbb{T}_2 = \mathbb{Z}$. Consider the problem*

$$\min\, (x_1-2)^2 + (x_2-2)^2 \quad \textit{subject to} \quad \begin{aligned} x_1^2 - x_2 &\leq 0, \\ 2x_1 + x_2 &\leq 8. \end{aligned} \tag{1.2}$$

Here

$$f(x_1,x_2) = (x_1-2)^2 + (x_2-2)^2, \quad (x_1,x_2) \in \mathbb{T}_1 \times \mathbb{T}_2,$$

$I_1 = \emptyset$, $I_2 = \{1,2\}$,

$$\begin{aligned} c_1(x) &= -x_1^2 + x_2, \\ c_2(x) &= -2x_1 - x_2 + 8, \quad (x_1,x_2) \in \mathbb{T}_1 \times \mathbb{T}_2. \end{aligned}$$

In Fig. 1.1 it is shown the contour of the function f which is the set of points for which f has a constant value. It is shown the set of points satisfying all the constraints, called feasible region, the point $x^* = (2,4)$ which is the solution of the problem (1.2)*. Note that the "infeasible side" of the inequality constraints is shaded.*

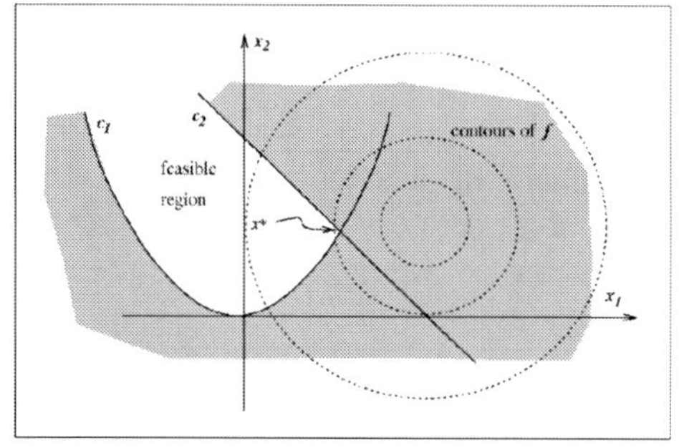

Figure 1.1. Geometrically representation of Problem 1.2.

Problems in the form (1.1) can be classified according to the nature of the objective function and constraints, the number of variables, the smoothness of the functions, and so on. An important distinction is between problems that have constraints on the variables and those that do not. Unconstrained optimization problems, for which we have $I_1 = I_2 = \emptyset$ in (1.1), arise directly in many practical problems. Unconstrained problems arise also as reformulations of constrained optimization problems, in which the constraints are replaced by penalization terms added to the objective function that have the effect of discouraging constraint violations. Constrained optimization problems arise from models in which constraints play an essential role. For instance, in imposing budgetary constraints in an economic problem or shape constraints in a design problem.

When the objective function and all the constraints are linear functions of x, the problem is a linear programming problem. Problems of this type are most widely formulatedand solved of all optimization problems, particularly in management, financial or economic applications. Nonlinear programming problems, in which at least some of the constraints or the objective function are nonlinear functions, arise naturally in the physical sciences, engineering and economic sciences as well.

Many problems for nonlinear optimization problems see a local solution. They do not always find the global solution. Global solutions are needed in some applications, but for many problems they are difficult to recognize and even more difficult to locate.

The concept of convexity is fundamental in optimization. Many practical problems possess this property, which makes them easier to solve both in theory and practice. The term "convex" can be applied both to sets and to functions.

Definition 1 *A set $A \subset \mathbb{R}^n$ is said to be a convex set if the straight line segment connecting any two points in A lies entirely inside A. Formally, if $x, y \in A$, then*

$$\alpha x + (1-\alpha)y \in A \quad \textit{for any} \quad \alpha \in [0,1].$$

Definition 2 *The function f is said to be convex if its domain A is a convex set and if for any $x, y \in A$ the following property is satisfied*

$$f(\alpha x + (1-\alpha)y) \leq \alpha f(x) + (1-\alpha)f(y) \quad \textit{for any} \quad \alpha \in [0,1]. \tag{1.3}$$

We say that f is strictly convex if the inequality (1.3) *is strict whenever $x \neq y$ and α is in the open interval* $(0,1)$. *A function f is said to be (strictly) concave if $-f$ is (strictly) convex.*

If the objective function in the optimization problem (1.1) and the feasible region are both convex, then any local solution of the problem (1.1) is a global solution.

In unconstrained optimization, we minimize an objective function that depends on time scale variables, with no restrictions at all on the values of these variables. The rigorous mathematical formulation is

$$\min_{x \in \Lambda^n} f(x),$$

where $f : \Lambda^n \to \mathbb{R}$ is a smooth function. Generally, we would like to find a global minimizer of f, a point where the function attains its least value. A formal definition is as follows.

Definition 3 *A point x^* is a global minimizer of f if*

$$f(x^*) \leq f(x) \quad \textit{for any} \quad x \in \Lambda^n.$$

The global minimizer can be difficult to find because our knowledge of f is usually only local. Obviously, we do not have a good picture of the overall shape of f. Many algorithms are able to find only a local minimizer, which is a point that achieves the smallest value of f in a neighbourhood. A formal definition is as follows.

Definition 4 *A point x^* is a local minimizer if there is a neighbourhood U of x^* such that*

$$f(x^*) \leq f(x) \quad \textit{for any} \quad x \in U, \quad \textit{and} \quad f(\sigma(x^*)) \geq f(x^*), \quad f(\rho(x^*)) \geq f(x^*). \tag{1.4}$$

If the inequalities (1.4) *are strict whenever $x \neq x^*$, we say that x^* is a strict local minimizer (or strong local minimizer). A point x^* is said to be an isolated local minimizer if there is a neighbourhood V of x^* such that x^* is the only local minimizer in V.*

While strict local minimizers are not always isolated, it is true that all isolated local minimizers are strict.

From the definitions given above, it might seem that the only way to find one whether a point x^* is a local minimum is to examine all the points in its immediate vicinity, to make sure that none of them has a smaller function value.

When the objective function is smooth, there are some efficient and practical ways to identify local minima. In particular, if f is twice continuously differentiable, we may be able to say that x^* is a local minimizer by examining the Δ-gradient

$$f^{\Delta}(x^*) = (f_{x_1}^{\Delta_1}(x^*), \ldots, f_{x_n}^{\Delta_n}(x^*)),$$

the ∇-gradient

$$f^{\nabla}(x^*) = (f_{x_1}^{\nabla_1}(x^*), \ldots, f_{x_n}^{\nabla_n}(x^*)),$$

the Δ-Hessian

$$H_{\Delta}(f, x, x^*) \quad = \quad \sum_{j=1}^{n} f_{x_j x_j}^{\Delta_j^2}(\xi^{0j}) h_2(x_j, x_j^*) + \sum_{j=2}^{n} f_{x_1 x_j}^{\Delta_1 \Delta_j}(\xi^{1j}) h_1(x_1, x_1^*) h_1(x_j, x_j^*)$$

$$+\sum_{j=3}^{n} f_{x_2x_j}^{\Delta_2\Delta_j}(\xi^{2j})h_1(x_2,x_2^*)h_1(x_j,x_j^*)$$

$$+\cdots$$

$$+f_{x_{n-1}x_n}^{\Delta_{n-1}\Delta_n}(\xi^{n-1j})h_1(x_{n_1},x_{n-1}^*)h_1(x,x_n^*),$$

and the ∇-Hessian

$$H_\nabla(f,x,x^*) = \sum_{j=1}^{n} f_{x_jx_j}^{\nabla_j^2}(\eta^{0j})H_2(x_j,x_j^*) + \sum_{j=2}^{n} f_{x_1x_j}^{\nabla_1\nabla_j}(\eta^{1j})h_1(x_1,x_1^*)h_1(x_j,x_j^*)$$

$$+\sum_{j=3}^{n} f_{x_2x_j}^{\nabla_2\nabla_j}(\eta^{2j})h_1(x_2,x_2^*)h_1(x_j,x_j^*)$$

$$+\cdots$$

$$+f_{x_{n-1}x_n}^{\nabla_{n-1}\nabla_n}(\eta^{n-1j})h_1(x_{n_1},x_{n-1}^*)h_1(x,x_n^*),$$

where ξ_l^{kj} and η_l^{kj} are between x_l and x_l^*, $l \in \{1,\ldots,n\}$, $k \in \{0,\ldots,n-1\}$, $j \in \{k+1,\ldots,n\}$, x is in a neighbourhood of x^*, and

$$h_1(x_j,x_j^*) = x_j - x_j^*,$$

$$h_2(x_j,x_j^*) = \int_{x_j^*}^{x_j} h_1(s,x_j^*)\Delta s,$$

$$H_2(x_j,x_j^*) = \int_{x_j^*}^{x_j} h_1(s,x_j^*)\nabla s.$$

The mathematical tool used the study minimizers of smooth functions are the Δ-Taylor theorem and ∇-Taylor theorem. These theorems are central to our analysis in this book. We state and prove the Δ-Taylor theorem.

Theorem 1 *(The Δ-Taylor Theorem) Let $a_j, b_j \in \mathbb{T}_j$, $a_j \le b_j$, $j \in \{1,\ldots,n\}$, and*

$$R = \{x \in \Lambda^n : a_j \le x_j \le b_j\}, \quad j \in \{1,\ldots,n\}.$$

Then, for any $f \in \mathscr{C}^2(R)$ *and for any* $x, x^0 \in R$ *there exist* ξ_l^{kj} *between* x_l *and* x_l^0, $l \in \{1,\ldots,n\}$, $k \in \{0,\ldots,n-1\}$, $j \in \{k+1,\ldots,n\}$, *such that*

$$f(x) = f(x^0) + \sum_{j=1}^{n} f_{x_j}^{\Delta_j}(x^0) + H_\Delta(f,x,x^0). \tag{1.5}$$

Proof 1 *Let*

$$\begin{aligned} x &= (x_1,\ldots,x_n), \\ x^0 &= (x_1^0,\ldots,x_n^0). \end{aligned}$$

Then, applying the time scale Taylor formula with respect to the first variable, we find that there exists a ξ_1^1 *between* x_1^0 *and* x_1 *such that*

$$\begin{aligned} f(x_1,x_2,\ldots,x_n) &= f(x_1^0,x_2,\ldots,x_n) + f_{x_1}^{\Delta_1}(x_1^0,x_2,\ldots,x_n)h_1(x_1,x_1^0) \\ &\quad + f_{x_1x_1}^{\Delta_1^2}(\xi_1^1,x_2,\ldots,x_n)h_2(x_1,x_1^0). \end{aligned} \tag{1.6}$$

Now, we apply the time scale Taylor formula for

$$f(x_1^0,x_2,\ldots,x_n) \quad \text{and} \quad f_{x_1}^{\Delta_1}(x_1^0,x_2,\ldots,x_n)$$

with respect to the second argument and we get that there are ξ_2^1 *and* ξ_2^2 *between* x_2 *and* x_2^0 *such that*

$$\begin{aligned} f(x_1^0,x_2,\ldots,x_n) &= f(x_1^0,x_2^0,\ldots,x_n) + f_{x_2}^{\Delta_2}(x_1^0,x_2^0,\ldots,x_n) \\ &\quad + f_{x_2x_2}^{\Delta_2^2}(x_1^0,\xi_2^1,\ldots,x_n)h_2(x_2,x_2^0) \end{aligned}$$

and

$$f_{x_1}^{\Delta_1}(x_1^0,x_2,\ldots,x_n) = f_{x_1}^{\Delta_1}(x_1^0,x_2^0,\ldots,x_n) + f_{x_1x_2}^{\Delta_1\Delta_2}(x_1^0,\xi_2^2,\ldots,x_n)h_1(x_2,x_2^0).$$

Now, employing (1.6), *we find*

$$\begin{aligned} f(x_1,x_2,\ldots,x_n) &= f(x_1^0,x_2^0,\ldots,x_n) + f_{x_2}^{\Delta_2}(x_1^0,x_2^0,\ldots,x_n) \\ &\quad + f_{x_2x_2}^{\Delta_2^2}(x_1^0,\xi_2^1,\ldots,x_n)h_2(x_2,x_2^0) \end{aligned}$$

$$+f_{x_1}^{\Delta_1}(x_1^0,x_2^0,\ldots,x_n)+f_{x_1x_2}^{\Delta_1\Delta_2}(x_1^0,\xi_2^2,\ldots,x_n)h_1(x_2,x_2^0)$$

$$+f_{x_1x_1}^{\Delta_1^2}(\xi_1^1,x_2,\ldots,x_n)h_2(x_1,x_1^0).$$

Now, we apply the time scale Taylor formula for

$$f(x_1^0,x_2^0,\ldots,x_n),\quad f_{x_1}^{\Delta_1}(x_1^0,x_2^0,\ldots,x_n)\quad \text{and}\quad f_{x_2}^{\Delta_2}(x_1^0,x_2^0,\ldots,x_n)$$

with respect to the third argument, and so on, while we get (1.5). *This completes the proof.*

As above, one can prove the ∇-Taylor theorem.

Theorem 2 *(The ∇-Taylor Theorem) Let $a_j,b_j\in\mathbb{T}_j$, $a_j\le b_j$, $j\in\{1,\ldots,n\}$, and*

$$R=\{x\in\Lambda^n : a_j\le x_j\le b_j\},\quad j\in\{1,\ldots,n\}.$$

Then, for any $f\in\mathscr{C}^2(R)$ and for any $x,x^0\in R$ there exist η_l^{kj} between x_l and x_l^0, $l\in\{1,\ldots,n\}$, $k\in\{0,\ldots,n-1\}$, $j\in\{k+1,\ldots,n\}$, such that

$$f(x)=f(x^0)+\sum_{j=1}^{n}f_{x_j}^{\nabla_j}(x^0)+H_\nabla(f,x,x^0).$$

For $x=(x_1,\ldots,x_n)$, $y=(y_1,\ldots,y_n)\in\Lambda^n$, we will write $x\le y$ if $x_j\le y_j$, $j\in\{1,\ldots,n\}$. We will write $x<y$ if $x_j\le y_j$ for any $j\in\{1,\ldots,n\}$, and there is at least one $l\in\{1,\ldots,n\}$ such that $x_l<y_l$.

Necessary conditions for optimality are derived assuming that x^* is a local minimizer and proving facts about the Δ-gradient $f^\Delta(x^*)$, the ∇-gradient $f^\nabla(x^*)$, the Δ-Hessian $H_\Delta(f,x,x^*)$ and the ∇-Hessian $H_\nabla(f,x,x^*)$.

Theorem 3 *(First Order Necessary Condition) Let $x^*\in\Lambda^n$ be a local minimizer for the objective function f and f be continuously differentiable in an open neighbourhood U of x^*. Then there are $\delta=(\delta_1,\ldots,\delta_n)>0$ and $\delta^1=(\delta_1^1,\ldots,\delta_n^1)>0$ such that*

$$f^\Delta(x)\ge 0,\quad x^*\le x<x^*+\delta,$$

and

$$f^\nabla(x)\le 0,\quad x^*-\delta^1<x\le x^*.$$

Proof 2 *Suppose the contrary, i.e., assume that*

$$f^{\Delta}(x) < 0, \quad x^* \le x < x^* + \delta,$$

or

$$f^{\nabla}(x) > 0, \quad x^* - \delta^1 < x \le x^*.$$

Then, there are $j, l \in \{1, \ldots, n\}$ *such that*

$$f_{x_j}^{\Delta_j}(x) < 0, \quad x^* \le x < x^* + \delta,$$

or

$$f_{x_l}^{\Delta_l}(x) > 0, \quad x^* - \delta^1 < x \le x^*.$$

Consider the following cases.

1. *Let* $\sigma_j(x_j^*) > x_j^*$. *Then*

$$\frac{f(x_1^*, \ldots, x_{j-1}^*, \sigma_j(x_j^*), x_{j+1}^*, \ldots, x_n^*) - f(x_1^*, \ldots, x_{j-1}^*, x_j^*, x_{j+1}^*, \ldots, x_n^*)}{\sigma_j(x_j^*) - x_j^*} = f_{x_j}^{\Delta_j}(x^*) < 0 \quad ,$$

whereupon

$$f(x_1^* \ldots, x_{j-1}^*, \sigma_j(x_j^*), x_{j+1}^*, \ldots, x_n^*) < f(x_1^*, \ldots, x_{j-1}^*, x_j^*, x_{j+1}^*, \ldots, x_n^*).$$

This is a contradiction.

2. *Let* $\sigma_j(x_j^*) = x_j^*$. *Then*

$$\begin{aligned} 0 &> f_{x_j}^{\Delta_j}(x^*) \\ &= \lim_{s_j \to x_j^*, s_j > x_j} \frac{f(x_1^*, \ldots, x_{j-1}^*, s_j, x_{j+1}^*, \ldots, x_n^*) - f(x_1^*, \ldots, x_{j-1}^*, x_j^*, x_{j+1}^*, \ldots, x_n^*)}{s_j - x_j^*}, \end{aligned}$$

whereupon

$$f(x_1^*, \ldots, x_{j-1}^*, s_j, x_{j+1}^*, \ldots, x_n^*) < f(x_1^*, \ldots, x_{j-1}^*, x_j^*, x_{j+1}^*, \ldots, x_n^*), \quad s_j > x_j^*.$$

This is a contradiction.

3. *Let $\rho_j(x_j^*) < x_j^*$. Then*

$$\frac{f(x_1^*,\ldots,x_{j-1}^*,x_j^*,x_{j+1}^*,\ldots,x_n^*) - f(x_1^*,\ldots,x_{j-1}^*,\rho_j(x_j^*),x_{j+1}^*,\ldots,x_n^*)}{x_j^* - \rho_j(x_j^*)} \quad = \quad f_{x_j}^{\nabla_j}(x^*)$$

$$> 0 \quad ,$$

whereupon

$$f(x_1^* \ldots,x_{j-1}^*,x_j^*,x_{j+1}^*,\ldots,x_n^*) > f(x_1^*,\ldots,x_{j-1}^*,\rho_j(x_j^*),x_{j+1}^*,\ldots,x_n^*).$$

This is a contradiction.

4. *Let $\rho_j(x_j^*) = x_j^*$. Then*

$$0 \quad < \quad f_{x_j}^{\nabla_j}(x^*)$$

$$= \quad \lim_{s_j \to x_j^*, s_j < x_j} \frac{f(x_1^*,\ldots,x_{j-1}^*,x_j^*,x_{j+1}^*,\ldots,x_n^*) - f(x_1^*,\ldots,x_{j-1}^*,s_j,x_{j+1}^*,\ldots,x_n^*)}{s_j - x_j^*},$$

whereupon

$$f(x_1^*,\ldots,x_{j-1}^*,s_j,x_{j+1}^*,\ldots,x_n^*) < f(x_1^*,\ldots,x_{j-1}^*,x_j^*,x_{j+1}^*,\ldots,x_n^*), \quad s_j < x_j^*.$$

This is a contradiction. This completes the proof.

Definition 5 *A point $x^* \in \Lambda^n$ is said to be a stationary point for the objective function f if*

$$f^{\Delta}(x^*) \geq 0 \quad \textit{and} \quad f^{\nabla}(x^*) \leq 0,$$

or

$$f^{\Delta}(x^*) \leq 0 \quad \textit{and} \quad f^{\nabla}(x^*) \geq 0.$$

By Theorem 3, it follows that any local minimizer is a stationary point.

Theorem 4 *(Second Order Necessary Conditions) Let x^* be a local minimizer of the objective function f and f be twice continuously differentiable in an open neighbourhood U of x^*. Then*

$$f^{\Delta}(x^*) \geq 0, \quad f^{\nabla}(x^*) \leq 0 \tag{1.7}$$

and

$$H_\Delta(f,x,x^*) \geq -\sum_{j=1}^{n} f_{x_j}^{\Delta_j}(x^*)h_1(x_j,x_j^*) \quad \textit{for} \quad x \in U, \tag{1.8}$$

and

$$H_\nabla(f,x,x^*) \geq -\sum_{j=1}^{n} f_{x_j}^{\nabla_j}(x^*)h_1(x_j,x_j^*) \quad \textit{for} \quad x \in U. \tag{1.9}$$

Proof 3 *By Theorem 3, it follows that* (1.7) *holds. Assume that* (1.8) *does not hold. Then there is a* $x^1 \leq x^*$, $x^1 \in U$, *such that*

$$H_\Delta(f,x^1,x^*) < -\sum_{j=1}^{n} f_{x_j}^{\Delta_j}(x^*)h_1(x_j^1,x_j^*).$$

By the Δ*-Taylor formula, we have*

$$f(x^1) = f(x^*) + \sum_{j=1}^{n} f_{x_j}^{\Delta_j}(x^*)h_1(x_j^1,x_j^*) + H_\Delta(f,x^1,x^*). \tag{1.10}$$

Therefore

$$f(x^1) < f(x^*).$$

This is a contradiction and then (1.8) *holds.*

Now, suppose that (1.9) *does not hold. Then there is a* $x^2 \geq x^*$, $x^2 \in U$, *such that*

$$H_\nabla(f,x^2,x^*) < -\sum_{j=1}^{n} f_{x_j}^{\nabla_j}(x^*)h_1(x_j^2,x_j^*).$$

Now, employing the ∇*-Taylor formula, we obtain*

$$f(x^2) = f(x^*) + \sum_{j=1}^{n} f_{x_j}^{\nabla_j}(x^*)h_1(x_j^2,x_j^*) + H_\nabla(f,x^2,x^*) \tag{1.11}$$

and then

$$f(x^2) < f(x^*).$$

This is a contradiction and then (1.9) *holds. This completes the proof.*

We now describe sufficient conditions, which are conditions on the derivatives of f at x^* that ensure that x^* is a local minimizer.

Theorem 5 *(Second Order Sufficient Conditions) Suppose that the objective function f is twice continuously differentiable in an open neighbourhood U of x^* that contains $\sigma(x^*)$ and $\rho(x^*)$, and*

$$f^{\Delta}(x^*) \geq 0, \quad H_{\Delta}(f,x,x^*) \geq 0, \quad x \geq x^*, \quad x \in U, \tag{1.12}$$

and

$$f^{\nabla}(x^*) \leq 0, \quad H_{\nabla}(f,x,x^*) \geq 0, \quad x \leq x^*, \quad x \in U. \tag{1.13}$$

Then x^* is a local minimizer of f.

Proof 4 *For $x \geq x^*$, $x \in U$, one has*

$$h_1(x_j,x_j^*) \geq 0, \quad j \in \{1,\ldots,n\},$$

and then employing the first inequality of (1.12), we find

$$\sum_{j=1}^{n} f_{x_j}^{\Delta_j}(x^*)h_1(x_j,x_j^*) \geq 0, \quad x \geq x^*, \quad x \in U.$$

Hence, the second inequality of (1.12) and the Δ-Taylor formula, we obtain

$$f(x) \geq f(x^*), \quad x \geq x^*, \quad x \in U.$$

In particular, $f(\sigma(x^*)) \geq f(x^*)$.
Now, for $x \leq x^*$, $x \in U$, one has

$$h_1(x_j,x_j^*) \leq 0, \quad j \in \{1,\ldots,n\},$$

and using the first inequality of (1.13), we arrive at

$$\sum_{j=1}^{n} f_{x_j}^{\nabla_j}(x^*)h_1(x_j,x_j^*) \geq 0, \quad x \leq x^*, \quad x \in U.$$

From this and the second inequality of (1.13), and the ∇-Taylor formula, we get

$$f(x) \geq f(x^*), \quad x \leq x^*, \quad x \in U,$$

and $f(\rho(x^*)) \geq f(x^*)$. This completes the proof.

When the objective function is convex, then the local and global minimizers are simple to characterize.

Theorem 6 *Let f be a convex function and x^* be its local minimizer. Then x^* is a global minimizer of f.*

Proof 5 *Assume that x^* is not a global minimizer of f. Then there is a $z \in \Lambda^n$ such that*

$$f(z) < f(x^*).$$

Consider the segment that joins x^* and z, i.e.,

$$x = \lambda z + (1-\lambda)x^*, \quad \lambda \in (0,1]. \tag{1.14}$$

Then, using that f is a convex function, we find

$$\begin{aligned} f(x) &= f(\lambda z + (1-\lambda)x^*) \\ &\leq \lambda f(z) + (1-\lambda) f(x^*) \\ &< \lambda f(x^*) + (1-\lambda) f(x^*) \\ &= f(x^*). \end{aligned}$$

Thus, any neighbourhood U of x^* contains a piece of the line segment (1.14) *and then for any neighbourhood V of x^* there is an $y \in V$ such that*

$$f(y) < f(x^*).$$

This is a contradiction because x^* is a local minimizer of f. Therefore x^* is a global minimizer of f. This completes the proof.

Chapter 2

Line Search Methods

In this chapter, we discuss some line search methods. The Wolfe conditions, strong Wolfe conditions and Goldstein conditions are introduced. Existence of step lengths that satisfy these conditions is proved. Closedness and convergence of line search methods are established. Rates of line search methods are investigated. The Newton method is introduced and some of its properties are deduced. A quasi Newton method is developed and it is proved that it is superlinearly convergent. They are described line search procedures based on interpolation of known function and derivative values of the function.

1. Description

A powerful collection of algorithms for unconstrained optimization of smooth functions are line search methods. We now give a broad description of their main properties.

Beginning at $x^0 \in \Lambda^n$ optimization methods generate a sequence of iterates $\{x^k\}_{k=0}^{\infty}$ that terminate when either no progress can be done or when it seems that a solution point has been approximated with sufficient accuracy. In deciding how to move from one iteration x^k to the next, the algorithms use information about the function f and possibly also information from earlier iterates $x^0, x^1, \ldots, x^{k-1}$. They use this information to find a new iterate x^{k+1} with a lower function value than x^k.

In the line search methods, we choose a direction p^k and search along this direction from the current iterate x^k for a new iterate with a lower function value. The distance to move along p^k can be found approximately solving the following minimization problem

$$\min_{\alpha>0} f(x^k+\alpha p^k), \quad x^{k+1}=x^k+\alpha p^k. \tag{2.1}$$

Here α is a positive scalar that belongs to a time scale $\mathbb{T}$, that contains 0, with forward jump operator, backward jump operator, delta differentiation operator and nabla differentiation operator $\widetilde{\sigma}$, $\widetilde{\rho}$, $\widetilde{\Delta}$ and $\widetilde{\nabla}$, $x^{k+1}, x^k \in \Lambda^n$, $p^k \in \mathbb{R}$. By solving (2.1) exactly, we derive the maximum benefit from the direction p^k. Below, we will use the notation

$$f_k = f(x^k), \quad k \in \mathbb{N}.$$

The steepest descent directions $-f_k^\Delta$ and $-f_k^\nabla$ are the most obvious choices for search directions for a line search method. It is intuitive, among all the directions we could move from x^k, it is the one along which f decreasing more rapidly. To see this, we appeal again to the Δ-Taylor theorem and ∇-Taylor theorem. The Δ-Taylor theorem tells us that for any search direction p and step-length parameter α, we have

$$\begin{aligned} f(x^{k+1}) &= f(x^k+\alpha p^k) \\ &= f(x^k)+\alpha p f^\Delta(x^k)^T + H_\Delta(f, x^{k+1}, x^k). \end{aligned}$$

The rate of change in f along the direction p at x^k is the coefficient of α, i.e., $pf^\Delta(x^k)^T$. Then the direction p of most rapid decrease is the solution of the minimizing problem

$$\min_p p f^\Delta(x^k)^T.$$

Since

$$p f^\Delta(x^k)^T = \|p\| \|f^\Delta(x^k)\| \cos\theta,$$

then the minimizer is attained when $\cos\theta = -1$ and

$$p = -f^\Delta(x^k)^T,$$

as claimed. As above, in the case of the ∇-Taylor formula, we get

$$p = -f^\nabla(x^k)^T.$$

Note that

$$H_\Delta(f, x^{k+1}, x^k) = H_\nabla(f, x^{k+1}, x^k) = O(\alpha^2),$$

and then

$$f(x^{k+1}) < f(x^k).$$

In computing the step-length α_k, we face a tradeoff. We would like to choose α_k to give a substantial reduction of f, but at the same time we do not want to speed too much time making the choice. The ideal choice would be the global minimization of the univariate function

$$\phi(\alpha) = f(x^k + \alpha p^k), \quad \alpha > 0,$$

but in general it is too expensive to identify this value. More practical strategies perform an inexact line search to identify a step-length that achieves adequate reductions in f at minimal cost.

2. The Time Scale Wolfe Conditions

Suppose that the objective function f is delta differentiable and continuously differentiable in the classical sense with $f^\Delta < 0$ and $f' < 0$. Here, with f' we denote the classical gradient of the objective function f.

An inexact line search condition stipulates that α_k should give sufficient decrease in the objective function f by the following inequality

$$f(x^k + \alpha_k p^k) \le f(x^k) + c_1 \alpha_k \left(\int_0^1 f'(x^k + h\widetilde{\sigma}(0)p^k)dh \right)^T p^k \tag{2.2}$$

for some $c_1 \in (0,1)$. In other words, the reduction in f should be proportional to both the step length α_k and the directional derivative

$$\left(\int_0^1 f'(x^k + h\widetilde{\sigma}(0)p^k)dh \right)^T p^k.$$

This sufficient decrease is illustrated in Fig. 2.1. The right hand side of the inequality (2.2) is denoted by

$$l(\alpha) = f(x^k) + c_1 \alpha \left(\int_0^1 f'(x^k + h\widetilde{\sigma}(0)p^k)dh \right)^T p^k.$$

The function l has negative slope

$$c_1 \left(\int_0^1 f'(x^k + h\widetilde{\sigma}(0)p^k)dh \right)^T p^k$$

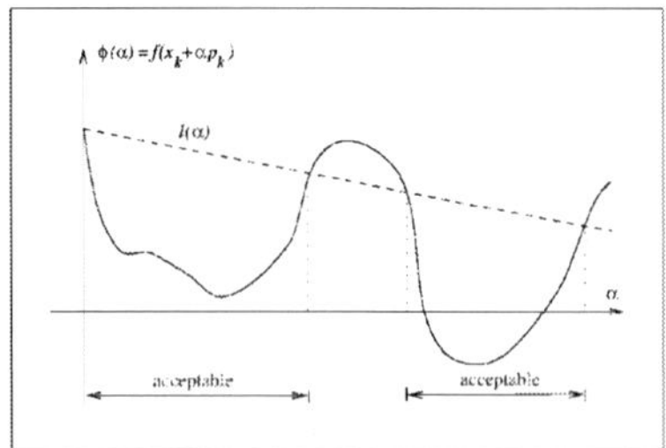

Figure 2.1. The sufficient decrease condition.

and since $c_1 \in (0,1)$ it lies above the graph of ϕ for small positive values of α. This sufficient decrease condition states that

$$\phi(\alpha) \leq l(\alpha).$$

The sufficient decrease condition is not enough to ensure that the algorithm makes reasonable progress because it is satisfied for all sufficiently small values of α. The rule out acceptable short steps, we introduce a second requirement, using the Pötzsche chain rule, and it is called the curvature condition and it requires α_k to satisfy

$$\begin{aligned} &\left(\int_0^1 f'(h(x^k+\widetilde{\sigma}(\alpha_k^1)p^k)+(1-h)(x^k+\alpha_k^1 p^k))dh\right)^T p^k \\ &\geq c_2\left(\int_0^1 f'(h(x^k+\widetilde{\sigma}(\alpha_k)p^k)+(1-h)(x^k+\alpha_k p^k))dh\right)^T p^k \end{aligned} \tag{2.3}$$

for some $\alpha_k^1 \in (0,\alpha_k)$ and $c_2 \in (c_1,1)$. The left hand side of the inequality (2.3) is the delta derivative of ϕ at α_k^1 and the right hand side is the delta derivative of ϕ at α_k. So, the curvature condition ensures that the slope of ϕ at α_k^1 is greater than c_2 times the slope of ϕ at some point α_k^1 between 0 and α_k. This makes sense because if the slope $\phi^{\Delta}(\alpha)$ is strongly negative, we have one indication that we can reduce f significantly by moving further along the chosen direction. The curvature condition is shown in Fig. 2.2. The sufficient decrease condition and the curvature

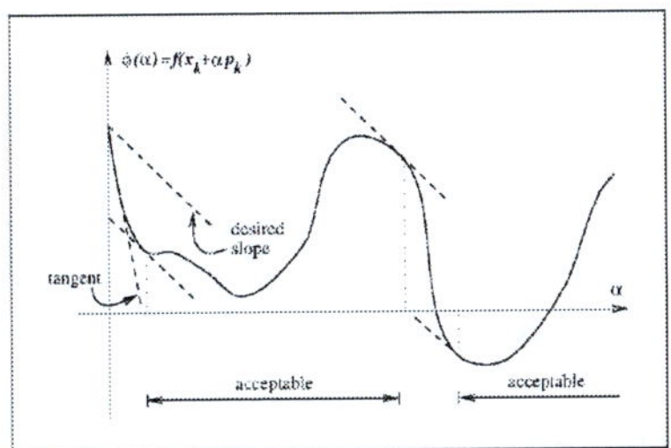

Figure 2.2. The curvature condition.

condition are known collectively as the time scale Wolfe conditions, shortly Wolfe conditions, and we restate them here for further reference.

$$f(x^k+\alpha_k p^k) \le f(x^k) + c_1\alpha_k \left(\int_0^1 f'(x^k + h\widetilde{\sigma}(0)p^k)dh\right)^T p^k$$

$$\left(\int_0^1 f'(h(x^k+\widetilde{\sigma}(\alpha_k^1)p^k) + (1-h)(x^k+\alpha_k^1 p^k))dh\right)^T p^k \tag{2.4}$$

$$\ge c_2 \left(\int_0^1 f'(x^k + h\widetilde{\sigma}(0)p^k)dh\right)^T p^k$$

for $0 < c_1 < c_2 < 1$ and for some α_k^1 between 0 and α_k. A step length may satisfy the Wolfe conditions without being particularly close to a minimizer of ϕ. However, we can modify the curvature condition to force α_k to lie at least a broad neighbourhood of a local minimizer or stationary points of ϕ. The strong time scale Wolfe conditions, shortly the strong Wolfe conditions, require α_k to satisfy

$$f(x^k+\alpha_k p^k) \le f(x^k)+c_1\alpha_k \left(\int_0^1 f'(x^k+h\tilde{\sigma}(0)p^k)dh\right)^T p^k$$

$$\left|\left(\int_0^1 f'(h(x^k+\tilde{\sigma}(\alpha_k^1)p^k)+(1-h)(x^k+\alpha_k^1 p^k))dh\right)^T p^k\right| \tag{2.5}$$

$$\le c_2\left|\left(\int_0^1 f'(x^k+h\tilde{\sigma}(0)p^k)dh\right)^T p^k\right|$$

for $0<c_1<c_2<1$ and for some $\alpha_k^1 \in (0,\alpha_k)$. The only difference with Wolfe conditions is that no longer allow the derivative $\phi^\Delta(\alpha_k)$ to be too positive.

Theorem 7 *Let the objective function f be delta differentiable and continuously differentiable in the classical sense with $f^\Delta<0$ and $f'<0$. Let also, f be bounded below along the ray*

$$\{x^k+\alpha p^k : \alpha>0\}. \tag{2.6}$$

If $0<c_1<c_2<1$, then there exist step lengths α^1 and α^2 that satisfy the Wolfe conditions (2.4) and the strong Wolfe conditions (2.5).

Proof 6 *Since f is bounded below the ray (2.6), we get that $\phi(\alpha)$ is bounded for any $\alpha>0$. Observe that l is unbounded below. Therefore l must interact ϕ at least once. Let α^1 be the smallest intersection value of α, i.e.,*

$$f(x^k+\alpha^1 p^k) = f(x^k)+\alpha^1 c_1\left(\int_0^1 f'(x^k+h\tilde{\sigma}(0)p^k)dh\right)^T p^k. \tag{2.7}$$

Thus, the first sufficient condition of (2.4) holds. On the other hand, by the mean value theorem, it follows that there is a $\alpha^2\in(0,\alpha^1)$ such that

$$f(x^k+\alpha^1 p^k)-f(x^k)\le \alpha^1\left(\int_0^1 f'(h(x^k+\tilde{\sigma}(\alpha^2)p^k)+(1-h)(x^k+\alpha^2 p^k))dh\right)^T p^k.$$

Hence, employing (2.7)*, we find*

$$\alpha^1 \left(\int_0^1 f'(h(x^k + \widetilde{\sigma}(\alpha^2)p^k) + (1-h)(x^k + \alpha^2 p^k))dh \right)^T p^k$$

$$\geq \alpha^1 c_1 \left(\int_0^1 f'(x^k + h\widetilde{\sigma}(0)p^k)dh \right)^T p^k,$$

or

$$\left(\int_0^1 f'(h(x^k + \widetilde{\sigma}(\alpha^2)p^k) + (1-h)(x^k + \alpha^2 p^k))dh \right)^T p^k$$

$$\geq \; c_1 \left(\int_0^1 f'(x^k + h\widetilde{\sigma}(0)p^k)dh \right)^T p^k$$

$$\geq \; c_2 \left(\int_0^1 f'(x^k + h\widetilde{\sigma}(0)p^k)dh \right)^T p^k,$$

i.e., α^1 *and* α^2 *satisfy the Wolfe conditions* (2.4)*. Because the terms in the last inequalities are negative, we conclude that* α^1 *and* α^2 *satisfy the strong Wolfe conditions* (2.5)*. This completes the proof.*

If $c_1 = c$ and $c_2 = 1 - c$ for $c \in \left(0, \frac{1}{2}\right)$, then the Wolfe conditions (2.4) are known as Goldstein conditions.

The sufficient decrease condition alone is not sufficient to ensure that the algorithm makes reasonable progress along the given search direction. If the line search algorithm chooses its candidate step lengths appropriately, by using the so called backtracking approach, we can dispense with the second condition in (2.4) and use just the sufficient decrease condition to terminate the line search procedure. Its basic form is as follows.

1. Choose $\overline{\alpha} > 0$, $\rho \in (0,1)$, $c \in (0,1)$. Set

$$\alpha \leftarrow \overline{\alpha}.$$

2. Repeat until

$$f(x^k+\alpha_k p^k) \le f(x^k)+c_1\alpha_k \left(\int_0^1 f'(x^k+h\widetilde{\sigma}(0)p^k)dh\right)^T p^k.$$

3. $\alpha \leftarrow \rho\alpha$.
4. Terminate with

$$\alpha_k = \alpha.$$

3. Closedness of Line Search Methods

Since searching along a line for a minimum point is a component part of most nonlinear algorithms, it is desirable to establish at once that this procedure is closed. This is the objective of this section.

To initiate a line search with respect to the objective function f, two vectors must be specified: the initial point $x^0 \in \Lambda^n$ and the direction d in which the search is to be made. The result of the search is a point.Thus, we define the search algorithm S as a mapping from $\Lambda^n \times \Lambda^n$ to Λ^n.

Assume that the search is to be made over the semi-infinite line emanating from x^0 in the direction d. We also assume that there is a minimum point along the line.

The mapping $S: \Lambda^n \times \Lambda^n \to \Lambda^n$ is defined in the following manner

$$S(x^0,d) = \{y : y = x^0+\alpha d \quad \text{for some} \quad \alpha \ge 0, \quad f(y) = \min_{\alpha\ge 0} f(x^0+\alpha d)\}.$$

In some cases there are many points y yielding the minimum. Thus, the mapping S is a set-valued mapping. We will verify that S is closed.

Theorem 8 *Let f be continuous on Λ^n. Then the mapping S is closed at (x^0,d).*

Proof 7 *Suppose that $\{x^k\}_{k\in\mathbb{N}}$ and $\{d^k\}_{k\in\mathbb{N}}$ are two sequences such that*

$$x^k \to x^0,$$

$$d^k \to d, \quad as \quad k\to\infty,$$

and $d \neq 0$. Assume also that $y^k \in S(x^k,d^k)$, $k\in\mathbb{N}$, and $y^k \to y$, as $k\to\infty$. We will show that $y \in S(x,d)$. Then, for any $k\in\mathbb{N}$, we have

$$y^k = x^k+\alpha_k d^k$$

for some $\alpha_k \in \mathbb{T}$. *Then*

$$y^k - x^k = \alpha_k d^k, \quad k \in \mathbb{N},$$

and

$$\alpha_k = \frac{\|y^k - x^k\|}{\|d^k\|}, \quad k \in \mathbb{N}.$$

Taking the limit on the right hand side of the above equality, we see that

$$\begin{aligned} \overline{\alpha} &= \lim_{k\to\infty} \alpha_k \\ &= \lim_{k\to\infty} \frac{\|y^k - x^k\|}{\|d^k\|} \\ &= \frac{\|y-x\|}{\|d\|}. \end{aligned}$$

Then,

$$y = x + \overline{\alpha} d.$$

For each $k \in \mathbb{N}$, *we have*

$$f(y^k) \le f(x^k + \alpha d^k).$$

Letting $k \to \infty$, *we obtain*

$$\begin{aligned} f(y) &= \lim_{k\to\infty} f(y^k) \\ &\le \lim_{k\to\infty} f(x^k + \alpha d^k) \\ &= f(x + \alpha d) \end{aligned}$$

and so,

$$f(y) \le \min_{\alpha \ge 0} f(x + \alpha d).$$

Hence, we conclude that $y \in S$. *This completes the proof.*

The requirement that $d \neq 0$ is natural. If $d = 0$, no search will be made. Theoretically, the map S can fail to be closed at $d = 0$. We will see this in the following example.

Example 2 *Let* $n = 1$ *and* $\mathbb{T}_1 = \mathbb{Z}$. *Let also, the objective function be the function*

$$f(x) = (x-1)^2, \quad x \in \mathbb{Z}.$$

We will show that $S(x,d)$ is not closed at $(0,0)$. Firstly, observe that for any $d > 0$, one has

$$\begin{aligned} \min_{\alpha \geq 0} f(\alpha d) &= \min_{\alpha \geq 0} (\alpha d - 1)^2 \\ &= f(1) \end{aligned}$$

and thus,

$$S(0,d) = 1.$$

On the other hand, we have

$$\min_{\alpha \geq 0} f(\alpha \cdot 0) = f(0)$$

and

$$S(0,0) = 0.$$

Thus, as $d \to 0$, we get

$$S(0,d) \not\to S(0,0).$$

4. Convergence of Line Search Methods

For convenience, we introduce the following notation

$$F_k(x^k, p^k, \alpha) = \int_0^1 f'(h(x^k + \tilde{\sigma}(\alpha)p^k) + (1-h)(x^k + \alpha p^k))dh.$$

To obtain global convergence, we should not only have well chosen step lengths but also well chosen search directions p^k. The angle θ_k between p^k and the steepest descent direction $-F_k$ is defined by

$$\cos\theta_k = \frac{-F_k^T p^k}{\|F_k\| \|p^k\|}.$$

The following theorem quantifies the effect of properly chosen step lengths and shows that the steepest descent method is globally convergent.

Theorem 9 *Consider any iteration of the form*

$$x^{k+1} = x^k + \alpha p^k,$$

where p^k is the descent direction and α_k, α_k^1 satisfy the Wolfe conditions (2.4)*. Let f be bounded below, delta differentiable and continuously differentiable with $f^\Delta < 0$ and $f' < 0$ in an open set U containing the level set*

$$\{x : f(x) \leq f(x^0)\},$$

where x^0 is the starting point of the iteration. Assume that F_k, $k \in \mathbb{N}$, satisfy the Lipschitz condition

$$\|F_k(x,p,\alpha) - F(y,q,\beta)\| \leq L\|(x+\alpha p) - (y+\beta q)\| \tag{2.8}$$

for some constant $L > 0$ and for any $x, y \in \Lambda^n$, $\alpha, \beta \in \mathbb{T}$ and $p, q \in \mathbb{R}^n$. Then

$$\sum_{k=0}^{\infty} (\cos\theta_k)^2 \|F_k\|^2 < \infty. \tag{2.9}$$

Proof 8 *Note that the conditions* (2.4) *can be written as follows*

$$f(x^{k+1}) \leq f(x^k) + c_1 \alpha_k F_k(x^k, p^k, \alpha)^T p^k$$

and

$$F_k(x^k, p^k, \alpha_k^1)^T p^k \geq c_2 F_k(x^k, p^k, 0)^T p^k.$$

Then, we have

$$\left(F_k(x^k, p^k, \alpha_k^1) - F_k(x^k, p^k, 0)\right)^T p^k \geq (c_2 - 1) F_k(x^k, p^k, 0)^T p^k.$$

On the other hand, applying the Lipschitz condition (2.8)*, we find*

$$\begin{aligned}(F_k(x^k, p^k, \alpha_k^1) - F_k(x^k, p^k, 0))^T p^k &\leq L\|x^k + \alpha_k^1 p^k - x^k\| \|p^k\| \\ &= L\alpha_k^1 \|p^k\|^2.\end{aligned}$$

Therefore

$$L\alpha_k^1 \|p^k\|^2 \geq (c_2 - 1) F_k(x^k, p^k, 0)^T p^k$$

and

$$\alpha_k^1 \geq \frac{c_2 - 1}{L} \frac{F_k(x^k, p^k, 0)^T p^k}{\|p^k\|^2}.$$

Hence, using that $\alpha_k^1 \in (0, \alpha_k)$, we find

$$\alpha_k \geq \frac{c_2 - 1}{L} \frac{F_k(x^k, p^k, 0)^T p^k}{\|p^k\|^2}.$$

By substituting this inequality in the first Wolfe inequality, we obtain

$$\begin{aligned} f_{k+1} &\leq f_k - c_1 \frac{1-c_2}{L} \frac{(F_k(x^k, p^k, 0)^T p^k)^2}{\|p^k\|^2} \\ &= f_k - c_1 \frac{1-c_2}{L} \frac{(F_k(x^k, p^k, 0)^T p^k)^2 \|F_k\|^2}{\|F^k\|^2 \|p^k\|^2} \\ &= f_k - c_1 \frac{1-c_2}{L} (\cos\theta_k)^2 \|F_k\|^2. \end{aligned}$$

Set

$$c = c_1 \frac{1-c_2}{L}$$

and we find

$$f_{k+1} \leq f_k - c(\cos\theta_k)^2 \|F_k\|^2.$$

By summing this expression over all indices less than or equal to k, we get

$$f_{k+1} \leq f_0 - c \sum_{j=0}^{k} (\cos\theta_j)^2 \|F_j\|^2,$$

or

$$c \sum_{j=0}^{k} (\cos\theta_j)^2 \|F_j\|^2 \leq f_0 - f_{k+1}.$$

Since f is bounded below, we conclude that there exists a constant $a > 0$ such that

$$c \sum_{j=0}^{k} (\cos\theta_j)^2 \|F_j\|^2 \leq a,$$

whereupon

$$\sum_{j=0}^{k} (\cos\theta_j)^2 \|F_j\|^2 \leq \frac{a}{c}.$$

Therefore

$$\sum_{j=0}^{\infty} (\cos\theta_j)^2 \|F_j\|^2 \leq \frac{a}{c}.$$

This completes the proof.

Similar results to this theorem hold when the Goldstein conditions or strong Wolfe conditions are used in place of the Wolfe conditions. For all these strategies, the step length selection implies the inequality (2.9). By the inequality (2.9), it follows that

$$(\cos\theta_k)^2\|F_k\|^2 \to 0, \quad \text{as} \quad k\to\infty.$$

If our method for choosing the search direction p^k in the iteration

$$x^{k+1} = x^k + \alpha p^k$$

ensures that θ_k is bounded away from $\frac{\pi}{2}$, then there is a positive constant $\delta > 0$ such that

$$\cos\theta_k \geq \delta > 0 \quad \text{for any} \quad k\in\mathbb{N},$$

and in this case we arrive at

$$\|F_k\| \to 0, \quad \text{as} \quad k\to\infty.$$

5. Rate of Convergence

It would seem that designing optimization algorithms with good convergence properties is easy, since all we need to ensure is that the search direction p^k does not tend to become orthogonal to the delta gradient f_k^Δ and nabla gradient f_k^∇, or that steepest descent steps are taken regularly. We could simply compute $\cos\theta_k$ at any iteration and turn p^k toward the steepest descent direction if $\cos\theta_k$ is smaller than some preselected constant $\delta > 0$. Angle tests of this type ensure global convergence, but they are undesirable. They may impede a fast rate of convergence, because for problems with an ill-conditional Δ-Hessian or ∇-Hessian, it may be necessary to produce search directions that are almost orthogonal to the delta gradient and nabla gradient. An inappropriate choice of the parameter δ may cause such steps to be rejected.

Algorithmic strategies that achieve rapid convergence can sometimes conflict with requirements of global convergence and vice versa. The challenge is to design algorithms that incorporate both properties: good global convergence guarantees and a rapid rate of convergence. We begin our study of convergence rates of line search methods by assuming that the Δ-Hessian satisfies the following conditions

$$\frac{a}{2}\|x^{k+1}-x^k\|^2 \leq H_\Delta(f,x^{k+1},x^k) \leq A f_k^{\Delta T} f_k^\Delta, \quad k\in\mathbb{N}, \tag{2.10}$$

where a and A are positive constants such that $\frac{1}{A}\in\mathbb{T}$. Assume that the line searches are exact $-f_k^\Delta$. Then, by (2.10) and the delta Taylor formula, we find

$$f\left(x^k-\alpha f_k^\Delta\right) \leq f(x^k) - \alpha f_k^{\Delta T} f_k^\Delta + \frac{A\alpha^2}{2} f_k^{\Delta T} f_k^\Delta.$$

Minimizing both sides separately with respect to α, the inequality will hold for the two minima. The minimum of the left hand side is $f(x^{k+1})$. The minimum of the right hand side occurs at $\alpha = \frac{1}{A}$. Thus, we find

$$\begin{aligned} f(x^{k+1}) &\leq f(x^k) - \frac{1}{2A} f_k^{\Delta T} f_k^{\Delta} \\ &= f(x^k) - \frac{1}{2A} \|f_k^{\Delta}\|^2. \end{aligned}$$

Let the optimal value be

$$f^* = f(x^*).$$

Subtracting the optimal value from both sides produces

$$f(x^{k+1}) - f^* \leq f(x^k) - f^* - \frac{1}{2A} \|f_k^{\Delta}\|^2. \tag{2.11}$$

Again applying the Δ-Taylor formula, we get

$$f(x) \geq f(x^k) + f_k^{\Delta T}(x - x^k) + \frac{a}{2} \|x - x^k\|^2.$$

Again we can minimize both sides separately. The minimum of the left hand side is the optimal solution value f^*. The minimum of the right hand side is

$$x^k - \frac{1}{a} f_k^{\Delta}.$$

Substituting, we get

$$f^* \geq f(x^k) - \frac{1}{2a} \|f_k^{\Delta}\|^2,$$

or

$$-\|f_k^{\Delta}\|^2 \leq 2a(f^* - f(x^k)).$$

From this and (2.11), we arrive at

$$\begin{aligned} f(x^{k+1}) - f^* &\leq f(x^k) - f^* + \frac{a}{A}(f^* - f(x^k)) \\ &= \left(1 - \frac{a}{A}\right)(f(x^k) - f^*). \end{aligned}$$

If we assume that

$$\frac{b}{2} \|x^{k+1} - x^k\|^2 \leq H_{\nabla}(f, x^{k+1}, x^k) \leq B f_k^{\nabla T} f_k^{\nabla}, \quad k \in \mathbb{N},$$

where b and B are positive constants such that $\frac{1}{B} \in \mathbb{T}$ and the line searches are exact $-f_k^{\nabla}$, $k \in \mathbb{N}$, as above, we obtain

$$f(x^{k+1}) - f^* \leq \left(1 - \frac{b}{B}\right)(f(x^k) - f^*).$$

This shows that the method of steepest descent makes progress even when it is not close to the solution.

6. The Newton Method

Now, we consider the Newton iteration for which the search is given by

$$p^k = -H_{\Delta}(f, x^{k+1}, x^k)^{-1} f_k^{\Delta}, \quad k \in \mathbb{N}. \tag{2.12}$$

Since the Δ-Hessian $H_{\Delta}(f, x^{k+1}, x^k)$, $k \in \mathbb{N}$, may not always be positive definite, p^k may not always be a descent direction, and many of the methods discussed in this chapter no longer apply.

Here, we discuss the local rate-of-convergence properties of the Newton method.

Theorem 10 *Let f be twice continuously delta differentiable function, $H_{\Delta}(f, x^{k+1}, x^k)$ be positive definite and Lipschitz continuous in a neighbourhood U of a solution x^* at which the sufficient conditions of Theorem 5 are satisfied and*

$$\|H_{\Delta}(f, x, y) - H_{\Delta}(f, x, z)\| \leq L\|y - z\|, \quad y, z \in U,$$

for some positive constant L. Consider the iteration

$$x^{k+1} = x^k + p^k, \quad k \in \mathbb{N},$$

where p^k, $k \in \mathbb{N}$, is given by (2.12). *Then, we have the following.*

1. *If the starting point x^0 is sufficiently close to x^*, then the sequence of iterations $\{x^k\}_{k \in \mathbb{N}}$ converges to x^*.*
2. *The rate of convergence of $\{x^k\}_{k \in \mathbb{N}}$ is quadratic.*
3. *The sequence $\{\|f_k^{\Delta}\|\}_{k \in \mathbb{N}}$ converges quadratically to zero.*

Proof 9 *From the definition of the Newton step and the optimality condition*

$$f^{\Delta*} \geq 0,$$

we get

$$\begin{aligned} x^k + p^k - x^* &= x^k - x^* - H_\Delta(f, x^{k+1}, x^k)^{-1} f_k^\Delta \\ &= H_\Delta(f, x^{k+1}, x^k)^{-1} \left(H_\Delta(f, x^{k+1}, x^k)(x^k - x^*) - f_k^\Delta \right) \\ &\leq H_\Delta(f, x^{k+1}, x^k)^{-1} \left(H_\Delta(f, x^{k+1}, x^k)(x^k - x^*) - (f_k^\Delta - f^{\Delta*}) \right), \quad k \in \mathbb{N}. \end{aligned}$$

Now, applying the mean value theorem, we conclude that there is a ξ^k between x^k and x^* such that

$$f_k^\Delta - f^{\Delta*} = H_\Delta(f, x^{k+1}, \xi^k)(x^k - x^*), \quad k \in \mathbb{N}.$$

Then

$$x^k + p^k - x^* \leq H_\Delta(f, x^{k+1}, x^k)^{-1} \left(H_\Delta(f, x^{k+1}, x^k) - H_\Delta(f, x^{k+1}, \xi^k) \right) (x^k - x^*), \quad k \in \mathbb{N},$$

and

$$\begin{aligned} \|x^k + p^k - x^*\| &\leq \|H_\Delta(f, x^{k+1}, x^k)^{-1}\| \|H_\Delta(f, x^{k+1}, x^k) - H_\Delta(f, x^{k+1}, \xi^k)\| \|x^k - x^*\| \\ &\leq L\|H_\Delta(f, x^{k+1}, x^k)^{-1}\| \|x^k - \xi^k\| \|x^k - x^*\| \\ &\leq L\|H_\Delta(f, x^{k+1}, x^k)^{-1}\| \|x^k - x^*\|^2, \quad k \in \mathbb{N}. \end{aligned}$$

Note that there is a constant L_1 such that

$$L\|H_\Delta(f, x^{k+1}, x^k)^{-1}\| \leq L_1, \quad k \in \mathbb{N}.$$

Therefore

$$\|x^k + p^k - x^*\| \leq L_1 \|x^k - x^*\|^2, \quad k \in \mathbb{N}.$$

Consequently the sequence $\{x^k\}_{k \in \mathbb{N}}$ converges to x^* and the rate of convergence is quadratic. Now, using that

$$p^k + H_\Delta(f, x^{k+1}, x^k)^{-1} f_k^\Delta = 0, \quad k \in \mathbb{N},$$

we find

$$f_k^\Delta + H_\Delta(f, x^{k+1}, x^k) p^k = 0, \quad k \in \mathbb{N}.$$

We apply the mean value theorem and we find that there is an η^k between x^k and x^{k+1} such that

$$f_{k+1}^{\Delta}-f_k^{\Delta}=H_{\Delta}(f,x^{k+1},\eta^k)(x^{k+1}-x^k),\quad k\in\mathbb{N}.$$

Hence,

$$\begin{aligned}
\|f_{k+1}^{\Delta}\| &= \|f_{k+1}^{\Delta}-f_k^{\Delta}-H_{\Delta}(f,x^{k+1},x^k)p^k\| \\
&= \|H_{\Delta}(f,x^{k+1},\eta^k)(x^{k+1}-x^k)-H_{\Delta}(f,x^{k+1},x^k)(x^{k+1}-x^k)\| \\
&= \|H_{\Delta}(f,x^{k+1},\eta^k)-H_{\Delta}(f,x^{k+1},x^k)\|\|p^k\| \\
&\le L\|\eta^k-x^k\|\|p^k\| \\
&\le L\|p^k\|^2 \\
&= L\|H_{\Delta}(f,x^{k+1},x^k)^{-1}\|^2\|f_k^{\Delta}\|^2,
\end{aligned}$$

proving that the delta gradient norms converge to zero quadratically. This completes the proof.

If we consider

$$p^k=-H_{\nabla}(f,x^{k+1},x^k)^{-1}f_k^{\nabla},\quad k\in\mathbb{N},\tag{2.13}$$

as above one can prove the following result

Theorem 11 *Let f be twice continuously delta differentiable function, $H_{\nabla}(f,x^{k+1},x^k)$ be positive definite and Lipschitz continuous in a neighbourhood U of a solution x^* at which the sufficient conditions of Theorem 5 are satisfied and*

$$\|H_{\nabla}(f,x,y)-H_{\nabla}(f,x,z)\|\le L\|y-z\|,\quad y,z\in U,$$

for some positive constant L. Consider the iteration

$$x^{k+1}=x^k+p^k,\quad k\in\mathbb{N},$$

where p^k, $k\in\mathbb{N}$, is given by (2.13). Then, we have the following.

1. *If the starting point x^0 is sufficiently close to x^*, then the sequence of iterations $\{x^k\}_{k\in\mathbb{N}}$ converges to x^*.*
2. *The rate of convergence of $\{x^k\}_{k\in\mathbb{N}}$ is quadratic.*
3. *The sequence $\{\|f_k^{\nabla}\|\}_{k\in\mathbb{N}}$ converges quadratically to zero.*

7. Quasi Newton Methods

Suppose that the search direction has the form

$$p^k = -B_k^{-1} f_k^{\Delta}, \quad k \in \mathbb{N}, \tag{2.14}$$

where B_k, $k \in \mathbb{N}$, are positive definite symmetric matrices that are updated at any iteration by a quasi Newton updating formula considered in the next chapters. In this case, we have the following result regarding the convergence rate.

Theorem 12 *Let f be twice continuously delta differentiable. Consider the iteration*

$$x^{k+1} = x^k + p^k, \quad k \in \mathbb{N},$$

where p^k, $k \in \mathbb{N}$, are given in (2.14). *Assume that the sequence $\{x^k\}_{k\in\mathbb{N}}$ converges to a point x^* for which one has*

$$f^{\Delta}(x^*) \geq 0 \quad \textit{and} \quad f^{\nabla}(x^*) \leq 0,$$

and $H_{\Delta}(f, x^{k+1}, x^k)$, $k \in \mathbb{N}$, are positive definite. Then the sequence $\{x^k\}_{k\in\mathbb{N}}$ converges superlinearly if and only if

$$\lim_{k\to\infty} \frac{\|(B_k - H_{\Delta}(f, x^{k+1}, x^k))p^k\|}{\|p^k\|} = 0. \tag{2.15}$$

Proof 10 *Let*

$$p^{kN} = -H_{\Delta}(f, x^{k+1}, x^k)^{-1} f_k^{\Delta}, \quad k \in \mathbb{N}.$$

Firstly, we will prove that the condition (2.15) *is equivalent to the condition*

$$p^k - p^{kN} = o(\|p^k\|), \quad k \in \mathbb{N}. \tag{2.16}$$

Really, suppose that (2.15) *holds. Then, we have*

$$\begin{aligned} p^k - p^{kN} &= p^k + H_{\Delta}(f, x^{k+1}, x^k)^{-1} f_k^{\Delta} \\ &= H_{\Delta}(f, x^{k+1}, x^k)^{-1} H_{\Delta}(f, x^{k+1}, x^k) p^k + H_{\Delta}(f, x^{k+1}, x^k)^{-1} f_k^{\Delta} \\ &= H_{\Delta}(f, x^{k+1}, x^k)^{-1} (H_{\Delta}(f, x^{k+1}, x^k) p^k + f_k^{\Delta}), \quad k \in \mathbb{N}. \end{aligned}$$

By (2.14), *we obtain*

$$f_k^{\Delta} = -B_k p^k, \quad k \in \mathbb{N}.$$

Therefore

$$\begin{aligned} p^k - p^{kN} &= H_\Delta(f,x^{k+1},x^k)^{-1}(H_\Delta(f,x^{k+1},x^k)p^k - B_k p^k) \\ &= H_\Delta(f,x^{k+1},x^k)^{-1}(H_\Delta(f,x^{k+1},x^k) - B_k)p^k \\ &= O\left(\|(H_\Delta(f,x^{k+1},x^k) - B_k)p^k\|\right) \\ &= o(\|p^k\|), \quad k \in \mathbb{N}. \end{aligned}$$

Conversely, assume that (2.16) *holds. We multiply it by* $H_\Delta(f,x^{k+1},x^k)$ *and we get*

$$\begin{aligned} H_\Delta(f,x^{k+1},x^k)o(\|p^k\|) &= H_\Delta(f,x^{k+1},x^k)(p^k - p^{kN}) \\ &= H_\Delta(f,x^{k+1},x^k)(-B_k^{-1}f_k^\Delta + H_\Delta(f,x^{k+1},x^k)^{-1}f_k^\Delta) \\ &= (B_k - H_\Delta(f,x^{k+1},x^k))B_k^{-1}f_k^\Delta \\ &= \left(H_\Delta(f,x^{k+1},x^k) - B_k\right)p^k, \quad k \in \mathbb{N}, \end{aligned}$$

whereupon we get (2.15)*. Next, we have*

$$\begin{aligned} \|x^k + p^k - x^*\| &\leq \|x^k + p^{kN} - x^*\| + \|p^{kN} - p^k\| \\ &= O(\|x^k - x^*\|^2) + o(\|p^k\|), \quad k \in \mathbb{N}. \end{aligned}$$

Note that

$$\|p^k\| = O(\|x^k - x^*\|), \quad k \in \mathbb{N}.$$

Therefore

$$\|x^k + p^k - x^*\| \leq o(\|x^k - x^*\|), \quad k \in \mathbb{N}.$$

This completes the proof.

If we take the search direction to be in the form

$$p^k = -B_k^{-1}f_k^\nabla, \quad k \in \mathbb{N}, \tag{2.17}$$

then, as above, we have the following result for the convergence rate.

Theorem 13 *Let f be twice continuously delta differentiable. Consider the iteration*

$$x^{k+1} = x^k + p^k, \quad k \in \mathbb{N},$$

where p^k, $k \in \mathbb{N}$, are given in (2.17). *Assume that the sequence $\{x^k\}_{k\in\mathbb{N}}$ converges to a point x^* for which one has*

$$f^{\Delta}(x^*) \geq 0 \quad \text{and} \quad f^{\nabla}(x^*) \leq 0,$$

and $H_{\nabla}(f, x^{k+1}, x^k)$, $k \in \mathbb{N}$, are positive definite. Then the sequence $\{x^k\}_{k\in\mathbb{N}}$ converges superlinearly if and only if

$$\lim_{k\to\infty} \frac{\|(B_k - H_{\nabla}(f, x^{k+1}, x^k))p^k\|}{\|p^k\|} = 0.$$

8. Interpolation

We begin by describing a line search procedure based on interpolation of known function.

Let $\mathscr{P}_n$, $n \in \mathbb{N}_0$, denote the set of all polynomials of degree $\leq n$ defined over the set $\mathbb{R}$ of real numbers. Let also, $n \in \mathbb{N}$ and $x_i \in \mathbb{T}$, $i \in \{0,1,\ldots,n\}$, be distinct and y_i, $i \in \{0,1,\ldots,n\}$, be given real numbers. We will find $p_n \in \mathscr{P}_n$ such that $p_n(x_i) = y_i$, $i \in \{0,1,\ldots,n\}$.

Theorem 14 *Suppose that $n \in \mathbb{N}$. Then there exist polynomials $L_k \in \mathscr{P}_n$, $k \in \{0,1,\ldots,n\}$, such that*

$$L_k(x_i) = \begin{cases} 1 & \text{if} \quad i = k \\ 0 & \text{if} \quad i \neq k, \end{cases}$$

$i,k \in \{0,1,\ldots,n\}$. Moreover,

$$p_n(x) = \sum_{k=0}^{n} L_k(x) y_k, \quad x \in \mathbb{T},$$

satisfies the condition $p_n(x_i) = y_i$, $i \in \{0,1,\ldots,n\}$, $p_n \in \mathscr{P}_n$.

Proof 11 *Define*

$$L_k(x) = C_k \sum_{i=0, i\neq k}^{n} (x - x_i), \quad x \in \mathbb{T},$$

where $C_k \in \mathbb{R}$, $k \in \{0, 1, \ldots, n\}$, *will be determined below. We have* $L_k(x_i) = 0$, $i \in \{0, 1, \ldots, n\}$, $i \neq k$, *and*

$$\begin{aligned} L_k(x_k) &= C_k \prod_{i=0, i\neq k}^{n} (x_k - x_i) \\ &= 1, \quad k \in \{0, 1, \ldots, n\}. \end{aligned}$$

Thus,

$$C_k = \frac{1}{\prod_{i=0, i\neq k}^{n} (x_k - x_i)}, \quad k \in \{0, 1, \ldots, n\},$$

and

$$L_k(x) = \prod_{i=0, i\neq k}^{n} \frac{x - x_i}{x_k - x_i}, \quad x \in \mathbb{T}, \quad k \in \{0, 1, \ldots, n\}. \tag{2.18}$$

We have that $L_k \in \mathscr{P}_n$, $k \in \{0, 1, \ldots, n\}$, *and* $p_n \in \mathscr{P}_n$. *This completes the proof.*

Theorem 15 *Assume that* $n \in \mathbb{N}_0$. *Let* $x_i \in \mathbb{T}$, $i \in \{0, 1, \ldots, n\}$, *be distinct and* $y_i \in \mathbb{R}$, $i \in \{0, 1, \ldots, n\}$. *Then there exists a unique polynomial* $p_n \in \mathscr{P}_n$ *such that*

$$p_n(x_i) = y_i, \quad i \in \{0, 1, \ldots, n\}.$$

Proof 12 *The existence of the polynomial* p_n *follows by Theorem 14. Suppose that there exist two polynomials* $p_n, q_n \in \mathscr{P}_n$ *such that*

$$p_n(x_i) = q_n(x_i) = y_i, \quad i \in \{0, 1, \ldots, n\}.$$

Then the polynomial $h_n = p_n - q_n$ *has* $n+1$ *distinct roots. Therefore* $h_n \equiv 0$ *or* $p_n \equiv q_n$. *This completes the proof.*

Definition 6 *Assume that* $n \in \mathbb{N}_0$. *Let* $x_i \in \mathbb{T}$, $i \in \{0, 1, \ldots, n\}$, *be distinct and* $y_i \in \mathbb{R}$, $i \in \{0, 1, \ldots, n\}$. *The polynomial*

$$p_n(x) = \sum_{k=0}^{n} L_k(x) y_k, \quad x \in \mathbb{T},$$

where L_k, $k \in \{0, 1, \ldots, n\}$, *are defined with* (2.18), *will be called the Lagrange interpolation polynomial of degree* n *with interpolation points* x_i, $i \in \{0, 1, \ldots, n\}$.

Definition 7 *Assume that $n \in \mathbb{N}_0$. Let $x_i \in [a,b] \subset \mathbb{T}$, $i \in \{0,1,\ldots,n\}$, be distinct and $f : [a,b] \to \mathbb{R}$ be a given function. The polynomial*

$$p_n(x) = \sum_{k=0}^{n} L_k(x) f(x_k), \quad x \in \mathbb{T},$$

where L_k, $k \in \{0,1,\ldots,n\}$, are defined with (2.18), will be called the Lagrange interpolation polynomial of degree n with interpolation points x_i, $i \in \{0,1,\ldots,n\}$, for the function f.

With $\mathscr{P}_n^{\widetilde{\sigma}}$, $n \in \mathbb{N}_0$, we will denote the set of all functions in the form

$$a_n(\widetilde{\sigma}(x))^n + a_{n-1}(\widetilde{\sigma}(x))^{n-1} + \cdots + a_1(x)\widetilde{\sigma}(x) + a_0, \quad x \in \mathbb{T},$$

where $a_j \in \mathbb{R}$, $j \in \{0,1,\ldots,n\}$. Let $a,b \in \mathbb{T}$, $a < b$. A function $g \in \mathscr{P}_n^{\widetilde{\sigma}}$ will be called a $\widetilde{\sigma}$-polynomial.

Definition 8 *Let $n \in \mathbb{N}_0$. The points $x_j \in [a,b)$, $j \in \{0,1,\ldots,n\}$, will be called $\widetilde{\sigma}$-distinct if $\widetilde{\sigma}(x_n) \leq b$ and*

$$\widetilde{\sigma}(x_0) < \widetilde{\sigma}(x_1) < \ldots < \widetilde{\sigma}(x_n).$$

As above, one can prove the following result.

Theorem 16 *Suppose that $n \in \mathbb{N}$ and $x_j \in \mathbb{T}$, $j \in \{0,1,\ldots,n\}$, are $\widetilde{\sigma}$-distinct. Then there exist unique $\widetilde{\sigma}$-polynomials $L_{\widetilde{\sigma}k} \in \mathscr{P}_n^{\widetilde{\sigma}}$, $k \in \{0,1,\ldots,n\}$, such that*

$$L_{\widetilde{\sigma}k}(x_i) = \begin{cases} 1 & \text{if} \quad i = k \\ 0 & \text{if} \quad i \neq k, \end{cases}$$

$i,k \in \{0,1,\ldots,n\}$. Moreover,

$$\begin{aligned} p_{\widetilde{\sigma}n}(x) &= \sum_{k=0}^{n} L_{\widetilde{\sigma}k}(x) y_k \\ &= \sum_{k=0}^{n} \left(\prod_{j=0, j\neq k}^{n} \frac{\widetilde{\sigma}(x) - \widetilde{\sigma}(x_j)}{\widetilde{\sigma}(x_k) - \widetilde{\sigma}(x_j)} \right) y_k, \quad x \in \mathbb{T}, \end{aligned}$$

satisfies the condition $p_{\widetilde{\sigma}n}(x_i) = y_i$, $i \in \{0,1,\ldots,n\}$, $p_{\widetilde{\sigma}n} \in \mathscr{P}_n^{\widetilde{\sigma}}$.

Definition 9 *Assume that $n \in \mathbb{N}_0$. Let $x_i \in \mathbb{T}$, $i \in \{0,1,\ldots,n\}$, be $\widetilde{\sigma}$-distinct and $y_i \in \mathbb{R}$, $i \in \{0,1,\ldots,n\}$. The $\widetilde{\sigma}$-polynomial*

$$p_{\widetilde{\sigma}n}(x) = \sum_{k=0}^{n} L_{\widetilde{\sigma}k}(x) y_k, \quad x \in \mathbb{T},$$

where $L_{\widetilde{\sigma}k}$, $k \in \{0,1,\ldots,n\}$, are defined in Theorem 16, will be called the $\widetilde{\sigma}$-Lagrange interpolation polynomial of degree n with $\widetilde{\sigma}$-interpolation points x_i, $i \in \{0,1,\ldots,n\}$.

Definition 10 *Assume that $n \in \mathbb{N}_0$. Let $x_i \in [a,b] \subset \mathbb{T}$, $i \in \{0,1,\ldots,n\}$, be $\widetilde{\sigma}$-distinct and $f : [a,b] \to \mathbb{R}$ be a given function. The $\widetilde{\sigma}$-polynomial*

$$p_{\widetilde{\sigma}n}(x) = \sum_{k=0}^{n} L_{\widetilde{\sigma}k}(x) f(x_k), \quad x \in \mathbb{T},$$

where $L_{\widetilde{\sigma}k}$, $k \in \{0,1,\ldots,n\}$, are defined in Theorem 16, will be called the $\widetilde{\sigma}$-Lagrange interpolation polynomial of degree n with $\widetilde{\sigma}$-interpolation points x_i, $i \in \{0,1,\ldots,n\}$, for the function f.

For given $\widetilde{\sigma}$-distinct points x_j, $j \in \{0,1,\ldots,n\}$ and two sets of real numbers y_j, z_j, $j \in \{0,1,\ldots,n\}$, $n \in \mathbb{N}_0$, we need to find a polynomial $p_{2n+1} \in \mathscr{P}_{2n+1}$ satisfying the conditions

$$p_{2n+1}(x_j) = y_j, \quad p_{2n+1}^{\widetilde{\Delta}}(x_j) = z_j, \quad j \in \{0,1,\ldots,n\}.$$

The construction of such a polynomial is similar to that of the Lagrange interpolation polynomial and it is given in the following theorem.

Theorem 17 (Hermite Interpolation Theorem) *Let $n \in \mathbb{N}_0$ and let $a, b \in \mathbb{T}$, $a < b$, and $x_j \in \mathbb{T}$, $j \in \{0,1,\ldots,n\}$, be $\widetilde{\sigma}$-distinct and $x_j \neq \widetilde{\sigma}(x_k)$ for all $j,k \in \{0,1,\ldots,n\}$. Let also, $y_j, z_j \in \mathbb{R}$, $j \in \{0,1,\ldots,n\}$. Then there exists a unique polynomial $p_{2n+1} \in \mathscr{P}_{2n+1}$ such that*

$$\begin{aligned} p_{2n+1}(x_j) &= y_j, \\ p_{2n+1}^{\widetilde{\Delta}}(x_j) &= z_j, \quad j \in \{0,1,\ldots,n\}. \end{aligned} \tag{2.19}$$

Proof 13 *For $x \in [a,b]$, define the polynomial*

$$M_k(x) = \prod_{j=0, j\neq k}^{n} \frac{x - \widetilde{\sigma}(x_j)}{x_k - \widetilde{\sigma}(x_j)}, \quad k \in \{0,1,\ldots,n\}.$$

We have

$$\begin{aligned} M_k(\widetilde{\sigma}(x_j)) &= 0, \\ M_k(x_k) &= 1, \quad j,k \in \{0,1,\ldots,n\}, \quad j \neq k. \end{aligned}$$

We will search a polynomial $p_{2n+1} \in \mathscr{P}_{2n+1}$ in the following form

$$p_{2n+1}(x) = \sum_{j=0}^{n} (y_j + (x - x_j)(\alpha_j y_j + \beta_j z_j)) M_j(x) L_j(x), \quad x \in [a,b],$$

where α_j, $\beta_j \in \mathbb{R}$, $j \in \{0,1,\ldots,n\}$, will be determined by the conditions (2.19)*. We have*

$$\begin{aligned} p_{2n+1}(x_k) &= \sum_{j=0}^{n} \big(y_j + (x_k - x_j)(\alpha_j y_j + \beta_j z_j)\big) M_j(x_k) L_j(x_k) \\ &= y_k, \\ p_{2n+1}^{\widetilde{\Delta}}(x) &= \sum_{j=0}^{n} (\alpha_j y_j + \beta_j z_j) M_j(\widetilde{\sigma}(x)) L_j(\widetilde{\sigma}(x)) \\ &\quad + \sum_{j=0}^{n} (y_j + (x - x_j)(\alpha_j y_j + \beta_j z_j)) \left(M_j^{\widetilde{\Delta}}(x) L_j(x) + M_j(\widetilde{\sigma}(x)) L_j^{\widetilde{\Delta}}(x)\right), \\ p_{2n+1}^{\widetilde{\Delta}}(x_k) &= \sum_{j=0}^{n} (\alpha_j y_j + \beta_j z_j) M_j(\widetilde{\sigma}(x_k)) L_j(\widetilde{\sigma}(x_k)) \\ &\quad + \sum_{j=0}^{n} (y_j + (x_k - x_j)(\alpha_j y_j + \beta_j z_j)) \left(M_j^{\widetilde{\Delta}}(x_k) L_j(x_k) + M_j(\widetilde{\sigma}(x_k)) L_j^{\widetilde{\Delta}}(x_k)\right) \\ &= (\alpha_k y_k + \beta_k z_k) M_k(\widetilde{\sigma}(x_k)) L_k(\widetilde{\sigma}(x_k)) \\ &\quad + y_k \left(M_k^{\widetilde{\Delta}}(x_k) + M_k(\widetilde{\sigma}(x_k)) L_k^{\widetilde{\Delta}}(x_k)\right) \\ &= z_k \end{aligned}$$

or

$$(\alpha_k y_k + \beta_k z_k) M_k(\widetilde{\sigma}(x_k)) L_k(\widetilde{\sigma}(x_k)) = z_k - y_k \left(M_k^{\widetilde{\Delta}}(x_k) + M_k(\widetilde{\sigma}(x_k)) L_k^{\widetilde{\Delta}}(x_k)\right),$$

or

$$\alpha_k y_k + \beta_k z_k = \frac{z_k - y_k\left(M_k^{\widetilde{\Delta}}(x_k) + M_k(\widetilde{\sigma}(x_k))L_k^{\widetilde{\Delta}}(x_k)\right)}{M_k(\widetilde{\sigma}(x_k))L_k(\widetilde{\sigma}(x_k))},$$

and

$$\begin{aligned} p_{2n+1}(x) &= \sum_{j=0}^{n}\left(y_j + \frac{z_j - y_j\left(M_j^{\widetilde{\Delta}}(x_j) + M_j(\widetilde{\sigma}(x_j))L_j^{\widetilde{\Delta}}(x_j)\right)}{M_j(\widetilde{\sigma}(x_j))L_j(\widetilde{\sigma}(x_j))}(x - x_j)\right)M_j(x)L_j(x) \\ &= \sum_{j=0}^{n}\left(\left(1 - \frac{M_j^{\widetilde{\Delta}}(x_j) + M_j(\widetilde{\sigma}(x_j))L_j^{\widetilde{\Delta}}(x_j)}{M_j(\widetilde{\sigma}(x_j))L_j(\widetilde{\sigma}(x_j))}(x - x_j)\right)y_j\right. \\ &\quad \left. + \frac{z_j}{M_j(\widetilde{\sigma}(x_j))L_j(\widetilde{\sigma}(x_j))}(x - x_j)\right)M_j(x)L_j(x), \end{aligned}$$

$x \in [a,b]$. Now, suppose that there are two polynomials such that $p_{2n+1}, q_{2n+1} \in \mathscr{P}_{2n+1}$ and

$$\begin{aligned} p_{2n+1}(x_k) &= q_{2n+1}(x_k) \\ &= y_k, \\ p_{2n+1}^{\widetilde{\Delta}}(x_k) &= q_{2n+1}^{\widetilde{\Delta}}(x_k) \\ &= z_k, \quad k \in \{0, 1, \ldots, n\}. \end{aligned}$$

Let

$$h_{2n+1} = p_{2n+1} - q_{2n+1}.$$

Then $h_{2n+1} \in \mathscr{P}_{2n+1}$ and it has at least $2n+2$ zeros. Thus,

$$h_{2n+1} \equiv 0 \quad or \quad p_{2n+1} \equiv q_{2n+1} \quad on \quad [a,b].$$

This completes the proof.

Definition 11 *Let $n \in \mathbb{N}_0$ and let a, $b \in \mathbb{T}$, $a < b$, and $x_j \in \mathbb{T}$, $j \in \{0, 1, \ldots, n\}$, be $\widetilde{\sigma}$-distinct. Let also, $y_j, z_j \in \mathbb{R}$, $j \in \{0, 1, \ldots, n\}$. Then the polynomial*

$$p_{2n+1}(x) = \sum_{j=0}^{n}\left(\left(1 - \frac{M_j^{\widetilde{\Delta}}(x_j) + M_j(\widetilde{\sigma}(x_j))L_j^{\widetilde{\Delta}}(x_j)}{M_j(\widetilde{\sigma}(x_j))L_j(\widetilde{\sigma}(x_j))}(x - x_j)\right)y_j\right.$$

$$+\frac{z_j}{M_j(\widetilde{\sigma}(x_j))L_j(\widetilde{\sigma}(x_j))}(x-x_j)\Big)M_j(x)L_j(x),$$

$x\in[a,b]$, is called Hermite interpolation polynomial for the set of values given in

$$\{(x_j,y_j,z_j): j\in\{0,1,\ldots,n\}\}.$$

Now, we will introduce $\widetilde{\sigma}$-Hermite polynomials.

Theorem 18 *Let $n\in\mathbb{N}_0$ and let a, $b\in\mathbb{T}$, $a<b$, and $x_j\in\mathbb{T}$, $j\in\{0,1,\ldots,n\}$, be $\widetilde{\sigma}$-distinct, the forward jump operator $\widetilde{\sigma}$ be $\widetilde{\Delta}$-differentiable on $[a,b]$ and $\widetilde{\sigma}^{\widetilde{\Delta}}(x_j)\neq 0$, $j\in\{0,1,\ldots,n\}$. Let also, $y_j,z_j\in\mathbb{R}$, $j\in\{0,1,\ldots,n\}$. Then there exists a unique $\widetilde{\sigma}$-polynomial $p_{\widetilde{\sigma}2n+1}\in\mathscr{P}^{\widetilde{\sigma}}_{2n+1}$ such that*

$$\begin{aligned} p_{\widetilde{\sigma}2n+1}(x_j) &= y_j,\\ p^{\widetilde{\Delta}}_{\widetilde{\sigma}2n+1}(x_j) &= z_j, \quad j\in\{0,1,\ldots,n\}. \end{aligned}$$

Proof 14 *Let M_k, $k\in\{0,1,\ldots,n\}$, be the polynomials as in the proof of Theorem 17. We will find a polynomial $p_{\widetilde{\sigma}2n+1}\in\mathscr{P}^{\widetilde{\sigma}}_{2n+1}$ in the following manner.*

$$p_{\widetilde{\sigma}2n+1}(x)=\sum_{j=0}^{n}(y_j+(\widetilde{\sigma}(x)-\widetilde{\sigma}(x_j))(\alpha_j y_j+\beta_j z_j))M_j(x)L_j(x),\quad x\in[a,b],$$

where α_j, $\beta_j\in\mathbb{R}$, $j\in\{0,1,\ldots,n\}$, will be determined below. We have

$$\begin{aligned} p_{\widetilde{\sigma}2n+1}(x_k) &= \sum_{j=0}^{n}(y_j+(\widetilde{\sigma}(x_k)-\widetilde{\sigma}(x_j))(\alpha_j y_j+\beta_j z_j))M_j(x_k)L_j(x_k)\\ &= y_k,\\ p^{\widetilde{\Delta}}_{\widetilde{\sigma}2n+1}(x) &= \sum_{j=0}^{n}\widetilde{\sigma}^{\widetilde{\Delta}}(x)(\alpha_j y_j+\beta_j z_j)M_j(\widetilde{\sigma}(x))L_j(\widetilde{\sigma}(x))\\ &\quad+\sum_{j=0}^{n}(y_j+(\widetilde{\sigma}(x)-\widetilde{\sigma}(x_j))(\alpha_j y_j+\beta_j z_j))\left(M_j^{\widetilde{\Delta}}(x)L_j(x)+M_j(\widetilde{\sigma}(x))L_j^{\widetilde{\Delta}}(x)\right),\\ p^{\widetilde{\Delta}}_{\widetilde{\sigma}2n+1}(x_k) &= \sum_{j=0}^{n}\widetilde{\sigma}^{\widetilde{\Delta}}(x_k)(\alpha_j y_j+\beta_j z_j)M_j(\widetilde{\sigma}(x_k))L_j(\widetilde{\sigma}(x_k)) \end{aligned}$$

$$+\sum_{j=0}^{n}\left(y_j+(\widetilde{\sigma}(x_k)-\widetilde{\sigma}(x_j))(\alpha_j y_j+\beta_j z_j)\right)\left(M_j^{\widetilde{\Delta}}(x_k)L_j(x_k)+M_j(\widetilde{\sigma}(x_k))L_j^{\widetilde{\Delta}}(x_k)\right)$$

$$= \widetilde{\sigma}^{\widetilde{\Delta}}(x_k)(\alpha_k y_k+\beta_k z_k)M_k(\widetilde{\sigma}(x_k))L_k(\widetilde{\sigma}(x_k))$$

$$+y_k\left(M_k^{\widetilde{\Delta}}(x_k)+M_k(\widetilde{\sigma}(x_k))L_k^{\widetilde{\Delta}}(x_k)\right)$$

$$= z_k,$$

or

$$\widetilde{\sigma}^{\widetilde{\Delta}}(x_k)\,(\alpha_k y_k+\beta_k z_k)\,M_k(\widetilde{\sigma}(x_k))L_k(\widetilde{\sigma}(x_k)) = z_k - y_k\left(M_k^{\widetilde{\Delta}}(x_k)+M_k(\widetilde{\sigma}(x_k))L_k^{\widetilde{\Delta}}(x_k)\right),$$

or

$$\alpha_k y_k+\beta_k z_k = \frac{z_k - y_k\left(M_k^{\widetilde{\Delta}}(x_k)+M_k(\widetilde{\sigma}(x_k))L_k^{\widetilde{\Delta}}(x_k)\right)}{\widetilde{\sigma}^{\widetilde{\Delta}}(x_k)M_k(\widetilde{\sigma}(x_k))L_k(\widetilde{\sigma}(x_k))},$$

and

$$p_{\widetilde{\sigma}2n+1}(x) = \sum_{j=0}^{n}\left(y_j+\frac{z_j-y_j\left(M_j^{\widetilde{\Delta}}(x_j)+M_j(\widetilde{\sigma}(x_j))L_j^{\widetilde{\Delta}}(x_j)\right)}{\widetilde{\sigma}^{\widetilde{\Delta}}(x_j)M_j(\widetilde{\sigma}(x_j))L_j(\widetilde{\sigma}(x_j))}(\widetilde{\sigma}(x)-\widetilde{\sigma}(x_j))\right)M_j(x)L_j(x)$$

$$= \sum_{j=0}^{n}\left(\left(1-\frac{M_j^{\widetilde{\Delta}}(x_j)+M_j(\widetilde{\sigma}(x_j))L_j^{\widetilde{\Delta}}(x_j)}{\widetilde{\sigma}^{\widetilde{\Delta}}(x_j)M_j(\widetilde{\sigma}(x_j))L_j(\widetilde{\sigma}(x_j))}(\widetilde{\sigma}(x)-\widetilde{\sigma}(x_j))\right)y_j\right.$$

$$\left.+\frac{z_j}{\widetilde{\sigma}^{\widetilde{\Delta}}(x_j)M_j(\widetilde{\sigma}(x_j))L_j(\widetilde{\sigma}(x_j))}(\widetilde{\sigma}(x)-\widetilde{\sigma}(x_j))\right)M_j(x)L_j(x),$$

$x \in [a,b]$. *Now, suppose that there are two* $\widetilde{\sigma}$*-polynomials such that* $p_{\widetilde{\sigma}2n+1}, q_{\widetilde{\sigma}2n+1} \in \mathscr{P}_{2n+1}^{\widetilde{\sigma}}$ *and*

$$p_{\widetilde{\sigma}2n+1}(x_k) = q_{\widetilde{\sigma}2n+1}(x_k)$$

$$= y_k,$$

$$p_{\widetilde{\sigma}2n+1}^{\widetilde{\Delta}}(x_k) = q_{\widetilde{\sigma}2n+1}^{\widetilde{\Delta}}(x_k)$$

$$= z_k, \quad k \in \{0,1,\ldots,n\}.$$

Let

$$h_{\widetilde{\sigma}2n+1} = p_{\widetilde{\sigma}2n+1} - q_{\widetilde{\sigma}2n+1}.$$

Then $h_{\widetilde{\sigma}2n+1} \in \mathscr{P}^{\widetilde{\sigma}}_{2n+1}$ *and it has at least* $2n+2$ *zeros. Thus,*

$$h_{\widetilde{\sigma}2n+1} \equiv 0 \quad or \quad p_{\widetilde{\sigma}2n+1} \equiv q_{\widetilde{\sigma}2n+1} \quad on \quad [a,b].$$

This completes the proof.

Definition 12 *Let* $n \in \mathbb{N}_0$ *and let* $a, b \in \mathbb{T}$, $a < b$, *and* $x_j \in \mathbb{T}$, $j \in \{0,1,\ldots,n\}$, *be* $\widetilde{\sigma}$-*distinct, the forward jump operator* $\widetilde{\sigma}$ *be* $\widetilde{\Delta}$-*differentiable on* $[a,b]$ *and* $\widetilde{\sigma}^{\widetilde{\Delta}}(x_j) \neq 0$, $j \in \{0,1,\ldots,n\}$. *Let also,* $y_j, z_j \in \mathbb{R}$, $j \in \{0,1,\ldots,n\}$. *The polynomial*

$$\begin{aligned} p_{\widetilde{\sigma}2n+1}(x) &= \sum_{j=0}^{n}\Bigg(\left(1-\frac{M_j^{\widetilde{\Delta}}(x_j)+M_j(\widetilde{\sigma}(x_j))L_j^{\widetilde{\Delta}}(x_j)}{\widetilde{\sigma}^{\widetilde{\Delta}}(x_j)M_j(\widetilde{\sigma}(x_j))L_j(\widetilde{\sigma}(x_j))}(\widetilde{\sigma}(x)-\widetilde{\sigma}(x_j))\right)y_j \\ &\quad +\frac{z_j}{\widetilde{\sigma}^{\widetilde{\Delta}}(x_j)M_j(\widetilde{\sigma}(x_j))L_j(\widetilde{\sigma}(x_j))}(\widetilde{\sigma}(x)-\widetilde{\sigma}(x_j))\Bigg)M_j(x)L_j(x), \end{aligned}$$

$x \in [a,b]$, *will be called* $\widetilde{\sigma}$-*Hermite interpolation polynomial for the set* $\{(x_j, y_j, z_j) : j \in \{0,1,\ldots,n\}\}$.

Definition 13 *Let* $n \in \mathbb{N}_0$ *and let* $a, b \in \mathbb{T}$, $a < b$, *and* $x_j \in \mathbb{T}$, $j \in \{0,1,\ldots,n\}$, *be* $\widetilde{\sigma}$-*distinct, the forward jump operator* $\widetilde{\sigma}$ *be* $\widetilde{\Delta}$-*differentiable on* $[a,b]$ *and* $\widetilde{\sigma}^{\widetilde{\Delta}}(x_j) \neq 0$, $j \in \{0,1,\ldots,n\}$. *Let also,* $f : [a,b] \to \mathbb{R}$ *be* $\widetilde{\Delta}$-*differentiable. The polynomial*

$$\begin{aligned} p_{\widetilde{\sigma}2n+1}(x) &= \sum_{j=0}^{n}\Bigg(\left(1-\frac{M_j^{\widetilde{\Delta}}(x_j)+M_j(\widetilde{\sigma}(x_j))L_j^{\widetilde{\Delta}}(x_j)}{\widetilde{\sigma}^{\widetilde{\Delta}}(x_j)M_j(\widetilde{\sigma}(x_j))L_j(\widetilde{\sigma}(x_j))}(\widetilde{\sigma}(x)-\widetilde{\sigma}(x_j))\right)f(x_j) \\ &\quad +\frac{f^{\widetilde{\Delta}}(x_j)}{\widetilde{\sigma}^{\widetilde{\Delta}}(x_j)M_j(\widetilde{\sigma}(x_j))L_j(\widetilde{\sigma}(x_j))}(\widetilde{\sigma}(x)-\widetilde{\sigma}(x_j))\Bigg)M_j(x)L_j(x), \end{aligned}$$

$x \in [a,b]$, *will be called* $\widetilde{\sigma}$-*Hermite interpolation polynomial for the function* f.

Chapter 3

Conjugate Direction Methods

In this chapter, we introduce some conjugate direction methods. The conjugate direction theorem is proved. Some descent properties of the conjugate direction method are given. In particular, the expanding subspace theorem is deduced. The conjugate direction algorithm is described and a variant of the conjugate direction theorem is formulated and proved. The C-G method is introduced as a generalization of the conjugate direction method. Some bounds of convergence are deducted.

1. Conjugate Directions

Conjugate direction methods are among the most useful techniques for solving large linear systems of equations. They can be adapted to solve nonlinear optimization problems. The remarkable properties of both linear and nonlinear conjugate gradient methods will be described in this chapter.

Suppose that the objective function f is twice delta and nabla differentiable such that

$$f_{x_j x_j}^{\Delta_j^2} \geq 0 \quad \text{and} \quad f_{x_j x_j}^{\nabla_j^2} \geq 0, \quad j \in \{1, \ldots, n\}.$$

Note that

$$h_2(x_j^{k+1}, x_j^k) \leq \frac{(x_j^{k+1} - x_j^k)^2}{2}, \quad j \in \{1, \ldots, n\}, \quad k \in \mathbb{N}.$$

Then, for the Δ-Hessian we have the following estimate

$$H_\Delta(f, x^{k+1}, x^k) \ \leq \ \frac{1}{2} \sum_{j=1}^{n} f_{x_j x_j}^{\Delta_j^2}(\xi^{0j})(x_j^{k+1} - x_j^k)^2$$

$$+\sum_{j=2}^{n} f_{x_1x_j}^{\Delta_1\Delta_j}(\xi^{1j})(x_1^{k+1}-x_1^k)(x_j^{k+1}-x_j^k)$$

$$+\sum_{j=3}^{n} f_{x_2x_j}^{\Delta_2\Delta_j}(\xi^{2j})(x_2^{k+1}-x_2^k)(x_j^{k+1}-x_j^k)$$

$$+\cdots$$

$$+f_{x_{n-1}x_n}^{\Delta_{n-1}\Delta_n}(\xi^{n-1j})(x_{n-1}^{k+1}-x_{n-1}^k)(x_n^{k+1}-x_n^k), \quad k\in\mathbb{N}.$$

Here ξ_l^{mj} are between x_l^k and x_l^{k+1}, $l\in\{1,\ldots,n\}$, $m\in\{0,1,\ldots,n-1\}$, $j\in\{m+1,\ldots,n\}$. If we denote

$$f^{\Delta^2} = \begin{pmatrix} f_{x_1x_1}^{\Delta_1^2} & f_{x_1x_2}^{\Delta_1\Delta_2} & \cdots & f_{x_1x_n}^{\Delta_1\Delta_n} \\ f_{x_2x_1}^{\Delta_2\Delta_1} & f_{x_2x_2}^{\Delta_2^2} & \cdots & f_{x_2x_n}^{\Delta_2\Delta_n} \\ \vdots & \vdots & \vdots & \vdots \\ f_{x_nx_1}^{\Delta_n\Delta_1} & f_{x_nx_2}^{\Delta_n\Delta_2} & \cdots & f_{x_nx_n}^{\Delta_n^2} \end{pmatrix},$$

$$p = \begin{pmatrix} x_1^{k+1}-x_1^k \\ x_2^{k+1}-x_2^k \\ \vdots \\ x_n^{k+1}-x_n^k \end{pmatrix}, \quad k\in\mathbb{N},$$

then we get

$$H_\Delta(f,x^{k+1},x^k)\le \frac{1}{2}p^T f^{\Delta^2} p, \quad k\in\mathbb{N},$$

and applying the Δ-Taylor formula, we find

$$\begin{aligned} f(x^{k+1}) &\le f(x^k)+f^\Delta(x^k)^T p+\frac{1}{2}p^T f^{\Delta^2}(\xi)p \\ &= f_k+f_k^{\Delta T}p+\frac{1}{2}p^T f^{\Delta^2}(\xi)p, \quad k\in\mathbb{N}, \end{aligned}$$

for some ξ between x^{k+1} and x^k. As above, we have

$$f(x^{k+1}) \le f(x^k)+f^\nabla(x^k)^T p+\frac{1}{2}p^T f^{\nabla^2}(\eta)p$$

$$= \quad f_k + f_k^{\nabla T} p + \frac{1}{2} p^T f^{\nabla^2}(\eta) p, \quad k \in \mathbb{N},$$

for some η between x^{k+1} and x^k, where

$$f^{\nabla^2} = \begin{pmatrix} f_{x_1x_1}^{\nabla_1^2} & f_{x_1x_2}^{\nabla_1\nabla_2} & \dots & f_{x_1x_n}^{\nabla_1\nabla_n} \\ f_{x_2x_1}^{\nabla_2\nabla_1} & f_{x_2x_2}^{\nabla_2^2} & \dots & f_{x_2x_n}^{\nabla_2\nabla_n} \\ \vdots & \vdots & \vdots & \vdots \\ f_{x_nx_1}^{\nabla_n\nabla_1} & f_{x_nx_2}^{\nabla_n\nabla_2} & \dots & f_{x_nx_n}^{\nabla_n^2} \end{pmatrix}.$$

Conjugate direction methods are invented and analyzed for the problem

$$\text{minimize} \quad g(p) = \frac{1}{2} p^T B p - b^T p, \tag{3.1}$$

where B is a positive definite and symmetric matrix, $b \in \mathbb{R}^n$.

Definition 14 *Two vectors x^1 and x^2 of $\mathbb{R}^n$ are said to be B-orthogonal or conjugate with respect to B if*

$$x^{1T} B x^2 = 0.$$

If $B = I$, then conjugacy is equivalent to the usual notion of orthogonality. A finite set of vectors $x^1, \dots, x^m$ are said to be B-orthogonal set if

$$x^{jT} B x^k = 0, \quad j \neq k, \quad j,k \in \{1, \dots, m\}.$$

An important property of B-orthogonal sets reads as follows.

Theorem 19 *If $x^1, \dots, x^m$ are B-orthogonal, then they are linearly independent.*

Proof 15 *Assume that there are $\lambda_j \in \mathbb{R}$, $j \in \{1, \dots, m\}$, such that*

$$\lambda_1 x^1 + \dots + \lambda_m x^m = 0.$$

Multiplying by B and taking scalar product with x^{jT} for some $j \in \{1, \dots, m\}$, we find

$$\lambda_j x^{jT} B x^j = 0.$$

Since B is positive definite, we have

$$x^{jT} B x^j > 0.$$

Therefore $\lambda_j = 0$, $j \in \{1, \dots, m\}$. This completes the proof.

Consider the minimizing problem (3.1). We have

$$g^{\Delta}(p) = Bp - b$$

and the unique solution of the minimizing problem is

$$Bp = b. \tag{3.2}$$

Therefore the minimizing problem (3.1) is equivalent to the linear equation problem. Suppose that $p^0, p^1, \ldots, p^{n-1}$ are n nonzero B-orthogonal vectors. By Theorem 19, it follows that they are linearly independent. Therefore the unique solution p^* of the equation (3.2) can be expanded in the following form

$$p^* = \alpha_0 p^0 + \alpha_1 p^1 + \cdots + \alpha_{n-1} p^{n-1}$$

for some $\alpha_j \in \mathbb{R}$, $j \in \{0, 1, \ldots, n-1\}$. Multiplying the last equation by B and then taking the scalar product with p^{jT}, $j \in \{0, 1, \ldots, n-1\}$, yields

$$\alpha_j p^{jT} B p^j = p^{jT} B p^*, \quad j \in \{0, 1, \ldots, n-1\},$$

and

$$\alpha_j = \frac{p^{jT} B p^*}{p^{jT} B p^j}, \quad j \in \{0, 1, \ldots, n-1\}.$$

Therefore

$$p^* = \sum_{j=0}^{n-1} \frac{p^{jT} B p^*}{p^{jT} B p^j} p^j.$$

The expression for p^* can be considered to be the result of an iterative process of n steps where at jth step $\alpha_j p^j$ is added.

Theorem 20 *(Conjugate Direction Theorem) Let $\{d^k\}_{k=0}^{n-1}$ be a set of nonzero B-orthogonal vectors. For any $p^0 \in \mathbb{R}^n$ the sequence $\{p^k\}_{k\in\mathbb{N}}$ generating in the following manner*

$$p^{k+1} = p^k + \alpha_k d^k, \quad k \in \mathbb{N}, \tag{3.3}$$

where

$$\alpha_k = -\frac{(Bp^k - b)^T d^k}{d^{kT} B d^k}, \quad k \in \mathbb{N},$$

converges to the unique solution p^* of the equation

$$Bp = b$$

after n steps, i.e.,

$$p^n = p^*.$$

Proof 16 *Since $\{d^k\}_{k=0}^{n-1}$ are B-orthogonal, by Theorem 19, it follows that they are linearly independent. Then*

$$p^* - p^0 = \alpha_0 d^0 + \alpha_1 d^1 + \cdots + \alpha_{n-1} d^{n-1}$$

for some $\alpha_j \in \mathbb{R}$, $j \in \{0, 1, \ldots, n-1\}$. Multiplying this equation with B and then taking the scalar product with d^{jT}, $j \in \{0, 1, \ldots, n-1\}$, we find

$$d^{jT} B(p^* - p^0) = \alpha_j d^{jT} B d^j, \quad j \in \{0, 1, \ldots, n-1\},$$

or

$$\alpha_j = \frac{d^{jT} B(p^* - p^0)}{d^{jT} B d^j}, \quad j \in \{0, 1, \ldots, n-1\}. \tag{3.4}$$

Now, following the iterative procedure from p^0 up to p^k gives

$$\begin{aligned} p^k &= p^{k-1} + \alpha_{k-1} d^{k-1} \\ &= p^{k-2} + \alpha_{k-2} d^{k-2} + \alpha_{k-1} d^{k-1} \\ &\vdots \\ &= p^0 + \alpha_0 d^0 + \alpha_1 d^1 + \cdots + \alpha_{k-1} d^{k-1}, \quad k \in \mathbb{N}, \end{aligned}$$

or

$$p^k - p^0 = \alpha_0 d^0 + \alpha_1 d^1 + \cdots + \alpha_{k-1} d^{k-1}, \quad k \in \mathbb{N}.$$

Hence, using the B-orthogonality of d^k, $k \in \{0, 1, \ldots, n-1\}$, multiplying by B and taking the scalar product with d^{kT}, $k \in \{0, 1, \ldots, n-1\}$, we find

$$d^{kT} B(p^k - p^0) = 0, \quad k \in \{0, 1, \ldots, n-1\},$$

whereupon

$$d^{kT} B p^k = d^{kT} B p^0, \quad k \in \{0, 1, \ldots, n-1\}.$$

Substituting the last equality into (3.4), we obtain

$$\begin{aligned} \alpha_k &= \frac{d^{kT} B(p^* - p^0)}{d^{kT} B d^k} \\ &= \frac{d^{kT} B p^* - d^{kT} B p^0}{d^{kT} B d^k} \end{aligned}$$

$$
\begin{aligned}
&= \frac{d^{kT}Bp^* - d^{kT}Bp^k}{d^{kT}Bd^k} \\
&= \frac{d^{kT}B(p^* - p^k)}{d^{kT}Bd^k} \\
&= \frac{d^{kT}(Bp^* - Bp^k)}{d^{kT}Bd^k} \\
&= \frac{d^{kT}(b - Bp^k)}{d^{kT}Bd^k} \\
&= -\frac{d^{kT}(Bp^k - b)}{d^{kT}Bd^k} \\
&= -\frac{(Bp^k - b)^T d^k}{d^{kT}Bd^k}, \quad k \in \{0, 1, \ldots, n-1\}.
\end{aligned}
$$

This completes the proof.

2. Descent Properties of the Conjugate Direction Method

Let $\{d^0, d^1, \ldots, d^{n-1}\}$ be as in Theorem 20. With $\mathscr{B}_k$, $k \in \{0, 1, \ldots, n-1\}$, we will denote the subspace spanned by $\{d^0, d^1, \ldots, d^{k-1}\}$.

Theorem 21 *(Expanding Subspace Theorem) For each $p^0 \in \mathbb{R}^n$, the sequence*

$$
p^{k+1} = p^k + \alpha_k d^k, \quad k \in \{0, 1, \ldots, n-1\},
$$

where

$$
\alpha_k = -\frac{(Bp^k - b)^T d^k}{d^{kT}Bd^k}, \quad k \in \{0, 1, \ldots, n-1\},
$$

has the property that p^k minimizes the function g on the line

$$
p = p^{k-1} + \alpha d^{k-1}, \quad k \in \{0, 1, \ldots, n-1\},
$$

as well as on the linear variety

$$
p^0 + \mathscr{B}_k.
$$

Proof 17 *We will prove that* $Bp^k - b \perp \mathscr{B}_k$, $k \in \{0, 1, \ldots, n-1\}$. *For this aim, we will use induction. Since* $\mathscr{B}_0$ *is the empty set, the assertion is true for* $k = 0$. *Assume that* $Bp^k - b \perp \mathscr{B}_k$ *for some* $k \in \{0, 1, \ldots, n-1\}$. *We will prove that* $Bp^{k+1} - b \perp \mathscr{B}_{k+1}$. *We have*

$$\begin{aligned} Bp^{k+1} - b &= B(p^k + \alpha_k d^k) - b \\ &= Bp^k + \alpha_k Bd^k - b \\ &= (Bp^k - b) + \alpha_k Bd^k, \quad k \in \{0, 1, \ldots, n-1\}. \end{aligned}$$

Then

$$\begin{aligned} d^{kT}(Bp^{k+1} - b) &= d^{kT}(Bp^k - b) + \alpha_k d^{kT} Bd^k \\ &= d^{kT}(Bp^k - b) - \frac{(Bp^k - b)^T d^k}{d^{kT} Bd^k} d^{kT} Bd^k \\ &= d^{kT}(Bp^k - b) - (Bp^k - b)^T d^k \\ &= d^{kT}(Bp^k - b) - d^{kT}(Bp^k - b) \\ &= 0, \quad k \in \{0, 1, \ldots, n-1\}. \end{aligned}$$

Next, for $j < k$, $k \in \{0, 1, \ldots, n-1\}$, *using that* $(Bp^k - b) \perp \mathscr{B}_k$, *we have*

$$d^{jT}(Bp^k - b) = 0.$$

Because $\{d^0, d^1, \ldots, d^{n-1}\}$ *is a B-orthogonal system, we have that*

$$d^{jT} Bd^k = 0, \quad j \neq k, \quad j, k \in \{0, 1, \ldots, n-1\}.$$

Therefore

$$\begin{aligned} d^{jT}(Bp^{k+1} - b) &= d^{jT}((Bp^k - b) + \alpha_k Bd^k) \\ &= d^{jT}(Bp^k - b) + \alpha_k d^{jT} Bd^k \\ &= 0. \end{aligned}$$

Consequently

$$(Bp^{k+1} - b) \perp \mathscr{B}_{k+1}.$$

Now, by induction, we conclude that

$$(Bp^k - b) \perp \mathscr{B}_k \quad \textit{for any} \quad k \in \{0,1,\ldots,n-1\}.$$

Since the function g is strictly convex, we have that p^k, $k \in \{0,1,\ldots,n-1\}$, minimizes g on the linear variety

$$p^0 + \mathscr{B}_k, \quad k \in \{0,1,\ldots,n-1\}.$$

Because

$$p^{k-1} + \alpha d^{k-1} \subset p^0 + \mathscr{B}_k, \quad k \in \{0,1,\ldots,n-1\},$$

we have that p^k, $k \in \{0,1,\ldots,n-1\}$, minimizes f on the line $p^0 + \mathscr{B}_k$, $k \in \{0,1,\ldots,n-1\}$. This completes the proof.

Note that

$$\mathscr{B}_k \subset \mathscr{B}_{k+1}, \quad k \in \{0,1,\ldots,n-1\},$$

and since p^k, $k \in \{0,1,\ldots,n-1\}$, minimizes g over $p^0 + \mathscr{B}_k$, $k \in \{0,1,\ldots,n-1\}$, we conclude that p^n is the overall minimum of g.

3. The Conjugate Direction Method

The conjugate direction method is obtained by selecting the successive direction vectors. The directions are determined sequentially at each step of the iteration. At step k one evaluates the current negative direction vector and adds to it a linear combination of the previous direction vectors to obtain a new conjugate direction vector along with to move.

We will describe the conjugate direction algorithm. Starting at any point $p^0 \in \mathbb{R}^n$, define

$$d^0 = b - Bp^0 \tag{3.5}$$

and

$$p^{k+1} = p^k + \alpha_k d^k, \tag{3.6}$$

where

$$\begin{aligned} \alpha_k &= -\frac{(Bp^k - b)^T d^k}{d^{kT} B d^k}, \\ d^{k+1} &= -(Bp^{k+1} - b) + \beta_k d^k, \\ \beta^k &= \frac{(Bp^{k+1} - b)^T B d^k}{d^{kT} B d^k}, \quad k \in \{0,1,\ldots,n-1\}. \end{aligned} \tag{3.7}$$

In this algorithm, the first step is identical to a steepest descent step and each succeeding step moves in a direction that is a linear combination of the current direction vector and the preceding direction vector.

To verify that the algorithm is conjugate direction algorithm, it is necessary to verify that the vectors $\{d^k\}_{k=0}^{n-1}$ are B-orthogonal. This is done in the next theorem, where the notation $[d^0, d^1, \ldots, d^k]$, $k \in \{0, 1, \ldots, n-1\}$, is used to denote the subspace spanned by the vectors $d^0, d^1, \ldots, d^k$.

Theorem 22 *(Conjugate Direction Theorem) The conjugate direction algorithm* (3.5)-(3.7) *is a conjugate direction method. If it does not terminate at p^k, then*

1. $$[Bp^0 - b, Bp^1 - b, \ldots, Bp^k - b] = [Bp^0 - b, B(Bp^0 - b), \ldots, B^k(Bp^0 - b)],$$

2. $$[d^0, d^1, \ldots, d^k] = [Bp^0 - b, B(Bp^0 - b), \ldots, B^k(Bp^0 - b)],$$

3. $d^{kT} Bd^j = 0$ *for* $j \le k-1$,

4. $$\alpha_k = \frac{(Bp^k - b)^T (Bp^k - b)}{d^{kT} Bd^k},$$

5. $$\beta_k = \frac{(Bp^{k+1} - b)^T (Bp^{k+1} - b)}{(Bp^k - b)^T (Bp^k - b)}, \quad k \in \{0, 1, \ldots, n-1\}.$$

Proof 18 *We will use induction. Note that the assertion is true for $k = 0$. Assume that the assertion is true for some $k \in \{0, 1, \ldots, n-2\}$. We will prove the assertion for $k+1$. We have*

$$Bp^{k+1} - b = Bp^k - b + \alpha_k Bd^k.$$

By the induction hypotheses, we have

$$Bp^k - b, \quad Bd^k \in [Bp^0 - b, B(Bp^0 - b), \ldots, B^{k+1}(Bp^0 - b)].$$

Thus,

$$Bp^{k+1} - b \in [Bp^0 - b, B(Bp^0 - b), \ldots, B^{k+1}(Bp^0 - b)].$$

Assume that

$$Bp^{k+1} - b \in [Bp^0 - b, B(Bp^0 - b), \ldots, B^k(Bp^0 - b)] = [d^0, d^1, \ldots, d^k].$$

Since

$$Bp^{k+1} - b \perp d^j, \quad j \in \{0, 1, \ldots, k\},$$

we get that

$$Bp^{k+1} - b = 0.$$

This is a contradiction. Therefore

$$Bp^{k+1} - b \notin [Bp^0 - b, B(Bp^0 - b), \ldots, B^k(Bp^0 - b)] = [d^0, d^1, \ldots, d^k].$$

Now, we consider

$$d^{k+1} = -(Bp^{k+1} - b) + \beta_k d^k.$$

Since

$$Bp^{k+1} - b, \quad d^k \in [Bp^0 - b, B(Bp^0 - b), \ldots, B^{k+1}(Bp^0 - b)],$$

we conclude that

$$d^{k+1} \in [Bp^0 - b, B(Bp^0 - b), \ldots, B^{k+1}(Bp^0 - b)]$$

and then

$$[d^0, d^1, \ldots, d^{k+1}] = [Bp^0 - b, B(Bp^0 - b), \ldots, B^{k+1}(Bp^0 - b)].$$

Next, observe that

$$\begin{aligned} d^{k+1T} Bd^j &= \left(-(Bp^{k+1} - b)^T + \beta_k d^{kT} \right) Bd^j \\ &= -(Bp^{k+1} - b)^T Bd^j + \beta_k d^{kT} Bd^j. \end{aligned}$$

If $j = k$, then

$$\begin{aligned} d^{k+1T} Bd^k &= -(Bp^{k+1} - b)^T Bd^k + \frac{(Bp^{k+1} - b)^T Bd^k}{d^{kT} Bd^k} d^{kT} Bd^k \\ &= -(Bp^{k+1} - b)^T Bd^k + (Bp^{k+1} - b)^T Bd^k \\ &= 0. \end{aligned}$$

Let $j < k$. Since $d^0, d^1, \ldots, d^{n-1}$ is a B-orthogonal system, we have that

$$d^{kT} Bd^j = 0.$$

Because

$$Bd^j \in [d^0, d^1, \ldots, d^{j+1}],$$

by the Expanding Subspace Theorem, we get that

$$(Bp^{k+1}-b)^T \perp Bd^j.$$

Therefore

$$d^{k+1T}Bd^j = 0.$$

By (3.7)*, we find*

$$d^k = -(Bp^k-b)+\beta_{k-1}d^{k-1},$$

whereupon taking scalar product with $(Bp^k-b)^T$*, we get*

$$\begin{aligned} -(Bp^k-b)^T d^k &= (Bp^k-b)^T(Bp^k-b)-\beta_{k-1}(Bp^k-b)^T d^{k-1} \\ &= (Bp^k-b)^T(Bp^k-b), \end{aligned}$$

where we have used the Expanding Subspace Theorem. Finally, note that

$$(Bp^{k+1}-b)^T(Bp^k-b) = 0$$

because $Bp^k-b \in [d^0,d^1,\ldots,d^k]$ *and*

$$Bp^{k+1}-b \perp [d^0,d^1,\ldots,d^k].$$

On the other hand,

$$p^{k+1} = p^k+\alpha_k d^k.$$

Then

$$p^{k+1}-p^k = \alpha_k d^k$$

and

$$Bp^{k+1}-Bp^k = \alpha_k Bd^k,$$

and

$$Bd^k = \frac{1}{\alpha_k}(Bp^{k+1}-Bp^k).$$

Now, taking scalar product with $(Bp^{k+1}-b)^T$*, we obtain*

$$\begin{aligned} (Bp^{k+1}-b)^T Bd^k &= \frac{1}{\alpha_k}(Bp^{k+1}-b)^T\left((Bp^{k+1}-b)-(Bp^k-b)\right) \\ &= \frac{1}{\alpha_k}\left((Bp^{k+1}-b)^T(Bp^{k+1}-b)-(Bp^{k+1}-b)^T(Bp^k-b)\right) \\ &= \frac{1}{\alpha_k}(Bp^{k+1}-b)^T(Bp^{k+1}-b) \end{aligned}$$

and

$$\alpha_k = \frac{(Bp^{k+1}-b)^T(Bp^{k+1}-b)}{(Bp^{k+1}-b)^T Bd^k}.$$

Thus, the assertion is true for $k+1$. By induction, we conclude that the assertion is true for any $k \in \{0,1,\ldots,n-1\}$. This completes the proof.

4. The C-G Method

The Conjugate Direction Theorem tells us that the spaces $\mathscr{B}_k$ over which we successively minimize are determined by original gradient $Bp^0 - b$ and multiplications of it by B. Each step of the method brings into consideration an additional power of B times $Np^0 - b$. Now, we will consider a new general approach for solving the minimization problem (3.1). Take a starting point p^0 and

$$p^{k+1} = p^0 + P_k(B)(Bp^0 - b), \tag{3.8}$$

where P_k is a polynomial of degree k. Selection of a set of coefficients for each polynomial P_k determines a sequence $\{p^k\}_{k\in\mathbb{N}}$. In fact, we have

$$\begin{aligned} p^{k+1} - p^* &= p^0 - p^* + P_k(B)(Bp^0 - b - Bp^* + b) \\ &= p^0 - p^* + P_k(B)B(p^0 - p^*) \\ &= (I + P_k(B)B)(p^0 - p^*). \end{aligned}$$

Define the function

$$E(p) = \frac{1}{2}(p - p^*)^T B(p - p^*).$$

Then

$$\begin{aligned} E(p^{k+1}) &= \frac{1}{2}(p^{k+1} - p^*)^T B(p^{k+1} - p^*) \\ &= \frac{1}{2}((I + P_k(B)B)(p^0 - p^*))^T B(I + P_k(B)B)(p^0 - p^*) \\ &= \frac{1}{2}(p^0 - p^*)^T (I + P_k(B)B)^T B(I + P_k(B)B)(p^0 - p^*) \\ &= \frac{1}{2}(p^0 - p^*)^T (I + P_k(B)B)B(I + P_k(B)B)(p^0 - p^*) \end{aligned}$$

$$= \frac{1}{2}(p^0 - p^*)^T B(I + P_k(B)B)^2(p^0 - p^*).$$

We may now pose the problem of selecting the polynomials P_k in such a way as to minimize $E(p^{k+1})$ with respect to the all polynomials of degree k. Expanding (3.8), we find

$$p^{k+1} = p^0 + \gamma_0(Bp^0 - b) + \gamma_1 B(Bp^0 - b) + \gamma_2 B^2(Bp^0 - b) + \cdots + \gamma_k B^k(Bp^0 - b),$$

where γ_j, $j \in \{0, 1, \ldots, k\}$, are the coefficients of the polynomial P_k. Because

$$\mathscr{B}_k = [d^0, d^1, \ldots, d^k] = [Bp^0 - b, B(Bp^0 - b), B^2(Bp^0 - b), \ldots, B^k(Bp^0 - b)],$$

the vector

$$p^{k+1} = p^0 + \alpha_0 d^0 + \alpha_1 d^1 + \cdots + \alpha_k d^k,$$

generated by the method of conjugate directions has precisely this form. By the Expanding Subspace Theorem, the coefficients γ_j, $j \in \{0, 1, \ldots, k\}$, determined by the conjugate direction process are such as to minimize $E(p^{k+1})$. Thus, the problem posed of selecting the optimal P_k is solved by the conjugate directions method.

Now, suppose that the vector $p^0 - p^*$ is written in the eigenvector expansion

$$p^0 - p^* = \delta_1 e_1 + \cdots + \delta_n e_n,$$

where $e_1, \ldots, e_n$ are the normalized eigenvectors of the matrix B. Since

$$\begin{aligned} B(p^0 - p^*) &= B(\delta_1 e_1 + \cdots + \delta_n e_n) \\ &= \delta_1 B e_1 + \cdots + \delta_n B e_n \\ &= \delta_1 \lambda_1 e_1 + \cdots + \delta_n \lambda_n e_n, \end{aligned}$$

where λ_j, $j \in \{1, \ldots, n\}$, are the eigenvalues of the matrix B, and since the eigenvectors of B are mutually orthogonal, we find

$$\begin{aligned} E(p^0) &= \frac{1}{2}(p^0 - p^*)^T B(p^0 - p^*) \\ &= \frac{1}{2}(\delta_1 e_1 + \cdots + \delta_n e_n)^T B(\delta_1 e_1 + \cdots + \delta_n e_n) \\ &= \frac{1}{2}\sum_{j=1}^{n} \delta_j^2 \lambda_j. \end{aligned}$$

Moreover,

$$\begin{aligned} E(p^{k+1}) &= \frac{1}{2}(p^0 - p^*)B(I + P_k(B)B)^2(p^0 - p^*) \\ &= \frac{1}{2}(\delta_1 e_1 + \cdots + \delta_n e_n)^T B(I + P_k(B)B)^2(\delta_1 e_1 + \cdots + \delta_n e_n) \\ &= \frac{1}{2}\sum_{j=1}^{n}(1 + P_k(\lambda_j)\lambda_j)^2 \lambda_j \delta_j^2 \\ &\leq \max_{\lambda_j}(1 + P_k(\lambda_j)\lambda_j)^2 \left(\frac{1}{2}\sum_{j=1}^{n}\lambda_j \delta_j^2\right) \\ &= \max_{\lambda_j}(1 + P_k(\lambda_j)\lambda_j)^2 E(p^0), \end{aligned}$$

i.e.,

$$E(p^{k+1}) \leq \max_{\lambda_j}(1 + P_k(\lambda_j)\lambda_j)^2 E(p^0).$$

Chapter 4

Quasi Newton Methods

In this chapter, we introduce the modified Newton method. We derive its convergence rate by employing the analysis of the method of steepest descent. We have algorithms for construction of the inverse of the Δ-Hessian in the case when it is a constant positive definite and symmetric matrix. We propose two procedures: a rank one procedure and a rank two procedure.

1. Modified Newton Method

Quasi Newton methods, line steepest descent method, require only the Δ-gradient and ∇-gradient of the objective function to be applied at each iterate. By measuring the changes in gradients, they construct a model of the objective function that is good enough to produce superlinear convergence. Moreover, since second delta and nabla derivatives are not required, quasi Newton methods are sometimes more efficient than the Newton method.

A basic iterative process for the problem

$$\text{minimize} \quad f(x)$$

is given by

$$x^{k+1} = x^k - \alpha_k S_k f_k^{\Delta T},$$

where S_k is an $n \times n$ matrix. If $S_k = I$, then we get the steepest descent method. If $S_k = H_\Delta(f, x^{k+1}, x^k)$, then we get the Newton method. Below suppose that S_k is a symmetric positive definite matrix. To derive the rate of convergence, consider the quadratic model

$$m(p) = \frac{1}{2} p^T B p - b^T p,$$

where B is a symmetric positive definite matrix. The algorithm becomes

$$p^{k+1} = p^k - \alpha_k S_k g^k, \tag{4.1}$$

where

$$\begin{aligned} g^k &= Bp^k - b, \\ \alpha_k &= \frac{g^{kT} S_k g^k}{g^{kT} S_k B S_k g^k}. \end{aligned} \tag{4.2}$$

We can derive the convergence rate of this algorithm by employing the analysis for the method of steepest descent. To do this, we have a need of the following auxiliary result, named the Kantorovich inequality.

Theorem 23 *(The Kantorovich Inequality) Let B be a positive definite symmetric matrix and λ and Λ be the smallest and largest eigenvalues of B, respectively. Then, for any $x \in \mathbb{R}^n$, one has*

$$\frac{(x^T x)^2}{(x^T Bx)(x^T B^{-1}x)} \geq \frac{4\lambda\Lambda}{(\lambda+\Lambda)^2}.$$

Proof 19 *Let the eigenvalues of B be*

$$0 < \lambda = \lambda_1 \leq \lambda_2 \leq \ldots \leq \lambda_n = \Lambda.$$

We change the coordinate system so that B becomes the diagonal matrix

$$B = \begin{pmatrix} \lambda_1 & 0 & \cdots & 0 \\ 0 & \lambda_2 & \cdots & 0 \\ \vdots & \vdots & \vdots & \vdots \\ 0 & 0 & \cdots & \lambda_n \end{pmatrix}.$$

In this coordinate system, we get

$$\frac{(x^T x)^2}{(x^T Bx)(x^T B^{-1}x)} = \frac{\left(\sum\limits_{j=1}^{n} x_j^2\right)}{\left(\sum\limits_{j=1}^{n} \lambda_j x_j^2\right)\left(\sum\limits_{j=1}^{n} \frac{1}{\lambda_j} x_j^2\right)}.$$

Set

$$\xi_j = \frac{x_j^2}{\sum\limits_{j=1}^{n} x_j^2}, \quad j \in \{1, \ldots, n\}.$$

Then

$$0 \le \xi_j \le 1, \quad j \in \{1, \ldots, n\},$$

and

$$\sum_{j=1}^{n} \xi_j = 1,$$

and

$$\begin{aligned}
\xi_j \lambda_j &= \frac{\lambda_j x_j^2}{\sum\limits_{j=1}^{n} x_j^2}, \\
\frac{\xi_j}{\lambda_j} &= \frac{\frac{1}{\lambda_j} x_j^2}{\sum\limits_{j=1}^{n} x_j^2}, \quad j \in \{1, \ldots, n\}.
\end{aligned}$$

Hence,

$$\begin{aligned}
\left(\sum_{j=1}^{n} \xi_j \lambda_j\right)\left(\sum_{j=1}^{n} x_j^2\right) &= \sum_{j=1}^{n} \lambda_j x_j^2, \\
\left(\sum_{j=1}^{n} \frac{\xi_j}{\lambda_j}\right)\left(\sum_{j=1}^{n} x_j^2\right) &= \sum_{j=1}^{n} \frac{1}{\lambda_j} x_j^2
\end{aligned}$$

and

$$\begin{aligned}
\frac{(x^T x)^2}{(x^T B x)(x^T B^{-1} x)} &= \frac{\left(\sum\limits_{j=1}^{n} x_j^2\right)^2}{\left(\sum\limits_{j=1}^{n} \lambda_j \xi_j\right)\left(\sum\limits_{j=1}^{n} \frac{\xi_j}{\lambda_j}\right)\left(\sum\limits_{j=1}^{n} x_j^2\right)^2} \\
&= \frac{\frac{1}{\sum\limits_{j=1}^{n} \lambda_j \xi_j}}{\sum\limits_{j=1}^{n} \frac{\xi_j}{\lambda_j}}.
\end{aligned}$$

Define

$$\phi_1(\xi) = \frac{1}{\sum\limits_{j=1}^{n} \xi_j \lambda_j},$$

$$\phi_2(\xi) = \sum_{j=1}^{n} \frac{\xi_j}{\lambda_j}.$$

Therefore

$$\frac{(x^T x)^2}{(x^T Bx)(x^T B^{-1}x)} = \frac{\phi_1(\xi)}{\phi_2(\xi)}.$$

Since $\sum_{j=1}^{n} \xi_j \lambda_j$ *is a point between* λ_1 *and* λ_n, *the value of* $\phi_1(\xi)$ *is a point on the curve* $\frac{1}{\lambda}$. *The value of* $\phi_2(\xi)$ *corresponds to a point that is under the graph of* $\frac{1}{\lambda}$ *and below the segment with end points* λ_1 *and* λ_n. *The minimum of the ratio*

$$\frac{\phi_1(\xi)}{\phi_2(\xi)}$$

is achieved for some

$$\lambda = \xi_1 \lambda_1 + \xi_n \lambda_n \quad \textit{with} \quad \xi_1 + \xi_n = 1.$$

Using that

$$\begin{aligned} \frac{\lambda_1 + \lambda_n - \xi_1 \lambda_1 - \xi_n \lambda_n}{\lambda_1 \lambda_n} &= \frac{\lambda_1 + \lambda_n - \xi_1 \lambda_1 - (1 - \xi_1)\lambda_n}{\lambda_1 \lambda_n} \\ &= \frac{\lambda_1 + \lambda_n - \xi_1 \lambda_1 - \lambda_n + \xi_1 \lambda_n}{\lambda_1 \lambda_n} \\ &= \frac{(1 - \xi_1)\lambda_1 + \xi_1 \lambda_n}{\lambda_1 \lambda_n} \\ &= \frac{\xi_n \lambda_1 + \xi_1 \lambda_n}{\lambda_1 \lambda_n} \\ &= \frac{\xi_1}{\lambda_1} + \frac{\xi_n}{\lambda_n}, \end{aligned}$$

we find that

$$\frac{(x^T x)^2}{(x^T Bx)(x^T B^{-1}x)} \geq \min_{\lambda_1 \leq \lambda \leq \lambda_n} \frac{\frac{1}{\lambda}}{\frac{\xi_1}{\lambda_1} + \frac{\xi_n}{\lambda_n}}$$

$$= \min_{\lambda_1 \le \lambda \le \lambda_n} \frac{\frac{1}{\lambda}}{\frac{\lambda_1+\lambda_n-\lambda}{\lambda_1\lambda_n}}$$

$$= \frac{4\lambda_1\lambda_n}{(\lambda_1+\lambda_n)^2}$$

$$= \frac{4\lambda\Lambda}{(\lambda+\Lambda)^2}.$$

This completes the proof.

Theorem 24 *Let p^* be the unique minimum of the function m and*

$$E(p) = \frac{1}{2}(p-p^*)^T B(p-p^*).$$

Then, for the algorithm (4.1), (4.2) *one has*

$$E(p^{k+1}) \le \left(\frac{\Lambda_k - \lambda_k}{\Lambda_k + \lambda_k}\right)^2 E(p^k),$$

where Λ_k and λ_k are the largest and smallest eigenvalue of S_kB, respectively.

Proof 20 *Firstly, note that*

$$E(p^k) - E(p^{k+1}) = \frac{1}{2}(p^k - p^*)^T B(p^k - p^*) - \frac{1}{2}(p^{k+1} - p^*)^T B(p^{k+1} - p^*)$$

$$= \frac{1}{2}\left((p^k-p^*)^T B(p^k-p^*) - (p^k - p^* - \alpha_k S_k g^k)^T B(p^k - p^* - \alpha_k S_k g^k)\right)$$

$$= \frac{1}{2}\Big((p^k-p^*)^T B(p^k-p^*) - (p^k-p^*)^T B(p^k - p^* - \alpha_k S_k g^k)$$

$$+\alpha_k (S_k g^k)^T B(p^k-p^*) - \alpha_k^2 (S_k g^k)^T B S_k g^k\Big)$$

$$= \frac{1}{2}\left(2\alpha_k (S_k g^k)^T B(p^k-p^*) - \alpha_k^2 (S_k g^k)^T B S_k g^k\right).$$

Note that

$$p^k - p^* = p^k - B^{-1}b$$

$$\begin{aligned} &= B^{-1}(Bp^k - b) \\ &= B^{-1}g^k. \end{aligned}$$

Then

$$\begin{aligned} E(p^k) - E(p^{k+1}) &= \frac{1}{2}\left(2\frac{g^{kT}S_kg^k}{g^{kT}S_kBS_kg^k}(S_kg^k)^TBB^{-1}g^k \right. \\ &\quad \left. - \frac{(g^{kT}S_kg^k)^2}{(g^{kT}S_kBS_kg^k)^2}g^{kT}S_kBS_kg^k\right) \\ &= \frac{1}{2}\left(2\frac{(g^{kT}S_kg^k)^2}{g^{kT}S_kBS_kg^k} - \frac{(g^{kT}S_kg^k)^2}{(g^{kT}S_kBS_kg^k)^2}g^{kT}S_kBS_kg^k\right) \\ &= \frac{1}{2}\left(2\frac{(g^{kT}S_kg^k)^2}{g^{kT}S_kBS_kg^k} - \frac{(g^{kT}S_kg^k)^2}{g^{kT}S_kBS_kg^k}\right) \\ &= \frac{1}{2}\frac{(g^{kT}S_kg^k)^2}{g^{kT}S_kBS_kg^k} \end{aligned}$$

and

$$\begin{aligned} E(p^k) &= \frac{1}{2}(p^k - p^*)^TB(p^k - p^*) \\ &= \frac{1}{2}(B^{-1}g^k)^TBB^{-1}g^k \\ &= \frac{1}{2}g^{kT}B^{-1}g^k. \end{aligned}$$

Thus,

$$\begin{aligned} \frac{E(p^k) - E(p^{k+1})}{E(p^k)} &= \frac{\frac{1}{2}\frac{(g^{kT}S_kg^k)^2}{g^{kT}S_kBS_kg^k}}{\frac{1}{2}g^{kT}B^{-1}g^k} \\ &= \frac{(g^{kT}S_kg^k)^2}{(g^{kT}S_kBS_kg^k)(g^{kT}B^{-1}g^k)}. \end{aligned}$$

Let

$$T_k = S_k^{\frac{1}{2}}BS_k^{\frac{1}{2}},$$

$$r^k = S_k^{\frac{1}{2}} g^k.$$

Then

$$S_k^{\frac{1}{2}} T_k S_k^{-\frac{1}{2}} = S_k B.$$

Therefore T_k and $S_k B$ are similar matrices and then they have the same eigenvalues. Now, applying the Kantorovich inequality, we get

$$\begin{aligned} \frac{E(p^k) - E(p^{k+1})}{E(p^k)} &= \frac{(r^{kT} r^k)^2}{(r^{kT} T_k r^k)(r^{kT} T_k^{-1} r^k)} \\ &\geq \frac{4\lambda_k \Lambda_k}{(\lambda_k + \Lambda_k)^2}, \end{aligned}$$

whereupon

$$E(p^k) - E(p^{k+1}) \geq \frac{4\lambda_k \Lambda_k}{(\lambda_k + \Lambda_k)^2} E(p^k),$$

or

$$\begin{aligned} E(p^{k+1}) &\leq \left(1 - \frac{4\lambda_k \Lambda_k}{(\lambda_k + \Lambda_k)^2}\right) E(p^k) \\ &= \frac{(\Lambda_k + \lambda_k)^2 - 4\lambda_k \Lambda_k}{(\lambda_k + \Lambda_k)^2} E(p^k) \\ &= \left(\frac{\Lambda_k - \lambda_k}{\Lambda_k + \lambda_k}\right)^2 E(p^k). \end{aligned}$$

This completes the proof.

2. Construction of the Inverse

Suppose that f is a function on Λ^n that is twice continuously delta and nabla differentiable. If $x^k, x^{k+1} \in \Lambda^n$ and

$$x^{k+1} = x^k + p^k,$$

applying the mean value theorem, we conclude that there is a ξ between x^k and x^{k+1} such that

$$f_{k+1}^{\Delta} - f_k^{\Delta} = H_{\Delta}(f, x^{k+1}, \xi) p^k.$$

Assume that the Δ-Hessian H_Δ is a constant F. Then

$$x^{k+1} - x^k = F p^k. \tag{4.3}$$

If n linearly independent directions $p^0, p^1, \dots, p^{n-1}$ and the corresponding

$$q^k = f_{k+1}^\Delta - f_k^\Delta$$

are known, then F is uniquely determined. Really, let P and Q be the $n \times n$ matrices with columns p^k and q^k, respectively. Then, we have

$$Q = FP$$

and hence,

$$F = QP^{-1}.$$

It is natural to attempt to construct successive approximations H_k of F^{-1} based on the data obtained from the first steps of a descent process in such a way that if F were constant the approximation would be construct with (4.3) for these steps. More precisely, if F were a constant H_{k+1} would satisfy

$$H_{k+1} q^j = p^j, \quad 0 < j \le k. \tag{4.4}$$

After n linearly independent steps we could then have $H_n = F^{-1}$.

Suppose that F is symmetric. Then, it is natural to require that H_k, the approximation to F^{-1}, is symmetric. We investigate the possibility of defining a recursion of the form

$$H_{k+1} = H_k + a_k z^k z^{kT} \tag{4.5}$$

which preserves symmetry. Setting j equal to k in (4.4) and substituting in (4.5), we obtain

$$\begin{aligned} p^k &= H_{k+1} q^k \\ &= (H_k + a_k z^k z^{kT}) q^k \\ &= H_k q^k + a_k z^k z^{kT} q^k. \end{aligned}$$

Taking the inner product with q^{kT}, we find

$$\begin{aligned} q^{kT} p^k &= q^{kT} H_k q^k + a_k q^{kT} z^k z^{kT} q^k \\ &= q^{kT} H_k q^k + a_k (z^{kT} q^k)^2, \end{aligned}$$

or

$$q^{kT}p^k - q^{kT}H_kq^k = a_k(z^{kT}q^k)^2. \tag{4.6}$$

By

$$p^k - H_kq^k = a_kz^kz^{kT}q^k,$$

we get

$$\begin{aligned}(p^k - H_kq^k)(p^k - H_kq^k)^T &= a_kz^kz^{kT}q^ka_k(z^kz^{kT}q^k)^T\\ &= a_k^2z^kz^{kT}q^kq^{kT}z^kz^{kT}\\ &= a_k^2z^k(z^{kT}q^k)^2z^{kT},\end{aligned}$$

whereupon

$$\frac{(p^k - H_kq^k)(p^k - H_kq^k)^T}{a_k(z^{kT}q^k)^2} = a_kz^kz^{kT}.$$

Therefore

$$\begin{aligned}H_{k+1} &= H_k + a_kz^kz^{kT}\\ &= H_k + \frac{(p^k - H_kq^k)(p^k - H_kq^k)^T}{a_k(z^{kT}q^k)^2},\end{aligned}$$

which in view of (4.6) leads finally to

$$\begin{aligned}H_{k+1} &= H_k + \frac{(p^k - H_kq^k)(p^k - H_kq^k)^T}{q^{kT}p^k - q^{kT}H_kq^k}\\ &= H_k + \frac{(p^k - H_kq^k)(p^k - H_kq^k)^T}{q^{kT}(p^k - H_kq^k)}.\end{aligned}$$

We have the following result.

Theorem 25 *Let F be a constant symmetric matrix and suppose that $p^0, p^1, \ldots, p^k$ are given vectors. Define the vectors*

$$q^j = Fp^j, \quad j \in \{0, 1, \ldots, k\}.$$

Starting with an initial matrix H_0, let

$$H_{j+1} = H_j + \frac{(p^j - H_jq^j)(p^j - H_jq^j)^T}{q^{jT}(p^j - H_jq^j)}, \quad j \in \{0, 1, \ldots, k\}.$$

Then

$$p_j = H_{k+1}q_j, \quad j \in \{0, 1, \ldots, k\}.$$

Proof 21 *We will use induction. For $k = 0$, the assertion is true. Assume that the assertion is true for some $k \in \mathbb{N}$. We will prove it for $k+1$. The relation was shown above to be true for H_{k+1}. For $j < k$, one has*

$$\begin{aligned} H_{k+1}q^j &= H_k q^j + \frac{(p^k - H_k q^k)(p^k - H_k q^k)^T}{q^{kT}(p^k - H_k q^k)} q^j \\ &= H_k q^j + \frac{p^k - H_k q^k}{q^{kT}(p^k - H_k q^k)} (p^{kT} q^j - q^{kT} H_k q^j). \end{aligned}$$

Set

$$y^k = \frac{p^k - H_k q^k}{q^{kT}(p^k - H_k q^k)}.$$

Then

$$H_{k+1}q^j = H_k q^j + y^k (p^{kT} q^j - q^{kT} H_k q^j).$$

On the other hand, by the induction hypothesis, we get

$$H_k q^j = p^j.$$

Then

$$H_{k+1}q^j = p^j + y^k (p^{kT} q^j - q^{kT} p^j). \tag{4.7}$$

Next, since

$$q^k = F p^k,$$

we find

$$q^{kT} = p^{kT} F.$$

Therefore

$$\begin{aligned} q^{kT} p^j &= p^{kT} F p^j \\ &= p^{kT} q^j. \end{aligned}$$

From this and (4.7)*, we arrive at*

$$H_{k+1}q^j = p^j.$$

Thus, the assertion is valid for $k+1$. By induction, we conclude that the assertion is true for any $k \in \mathbb{N}$. This completes the proof.

This scheme is known as a rank one correction.

3. A Rank Two Correction Procedure

A rank two correction procedure is as follows. Starting with any symmetric positive definite matrix H_0, any point x^0, and with $k = 0$.

1. Set
$$d^k = -H_k f_k^{\Delta}.$$
2. Minimize
$$f(x^k + \alpha d^k)$$
with respect to α to obtain x^{k+1},
$$p^k = \alpha_k d^k,$$
and f_{k+1}^{Δ}.
3. Set
$$q^k = f_{k+1}^{\Delta} - f_k^{\Delta}$$
and
$$H_{k+1} = H_k + \frac{p^k p^{kT}}{p^{kT} q^k} - \frac{H_k q^k q^{kT} H_k}{q^{kT} H_k q^k}.$$

We first demonstrate that if H_k is positive definite, then H_{k+1} is positive definite. Take $x \in \mathbb{R}^n$ arbitrarily so that $x \neq 0$. Then

$$x^T H_{k+1} x = x^T H_k x + \frac{(x^T p^k)^2}{p^{kT} q^k} - \frac{(x^T H_k q^k)^2}{q^{kT} H_k q^k}. \tag{4.8}$$

Define

$$\begin{aligned} a &= H_k^{\frac{1}{2}} x, \\ b &= H_k^{\frac{1}{2}} q^k. \end{aligned}$$

Then, we can rewrite (4.8) as follows

$$x^T H_{k+1} x = \frac{(a^T a)(b^T b) - (a^T b)^2}{b^T b} + \frac{(x^T p^k)^2}{p^{kT} q^k}.$$

Also, we have

$$p^{kT} q^k = p^{kT} f_{k+1}^{\Delta} - p^{kT} f_k^{\Delta}$$

$$= -p^{kT} f_k^{\Delta},$$

where we have used that

$$p^{kT} f_{k+1}^{\Delta} = 0$$

because x^{k+1} is the minimum point of f along p^k. By the definition of p^k, we find

$$\begin{aligned}
p^{kT} q^k &= \alpha_k d^{kT} q^k \\
&= -\alpha_k f_k^{\Delta T} H_k q^k \\
&= -\alpha_k f_k^{\Delta T} H_k (f_{k+1}^{\Delta} - f_k^{\Delta}) \\
&= -\alpha_k f_k^{\Delta T} H_k f_{k+1}^{\Delta} + \alpha_k f_k^{\Delta T} H_k f_k^{\Delta} \\
&= \alpha_k f_k^{\Delta T} H_k f_k^{\Delta}
\end{aligned}$$

and hence,

$$x^T H_{k+1} x = \frac{(a^T a)(b^T b) - (a^T b)^2}{b^T b} + \frac{(x^T p^k)^2}{\alpha_k f_k^{\Delta T} H_k f_k^{\Delta}}. \tag{4.9}$$

By the Cauchy-Schwartz inequality, we have

$$(a^T a)(b^T b) \geq (a^T b)^2,$$

and since H_k is positive definite, we conclude that

$$f_k^{\Delta T} H_k f_k^{\Delta} > 0.$$

Then

$$x^T H_{k+1} x \geq 0.$$

We will show that both terms in (4.9) do not both vanish simultaneously. The first term vanishes if and only if a and b are proportional. By the definition of a and b, this in turn implies that x and q^k are proportional, say

$$x = \beta q^k.$$

In that case, we find

$$p^{kT} x = \beta p^{kT} q^k$$

$$= \beta\alpha_k f_k^{\Delta T} H_k f_k^{\Delta}$$

$$\neq 0.$$

Therefore

$$x^T H_{k+1} x > 0$$

for all nonzero x and H_{k+1} is symmetric.

Theorem 26 *Assume that f is with constant Δ-Hessian F that is symmetric positive definite. Then*

$$p_i^T F p_j = 0, \quad 0 \leq i < j \leq k, \tag{4.10}$$

and

$$H_{k+1} F p_j = p_j, \quad 0 \leq j \leq k. \tag{4.11}$$

Proof 22 *We have*

$$\begin{aligned} q^k &= f_{k+1}^{\Delta} - f_k^{\Delta} \\ &= F(x^{k+1} - x^k) \\ &= F p^k \end{aligned}$$

and

$$\begin{aligned} H_{k+1} F p^k &= H_{k+1} q^k \\ &= \left(H_k + \frac{p^k p^{kT}}{p^{kT} q^k} - \frac{H_k q^k q^{kT} H_k}{q^{kT} H_k q^k} \right) q^k \\ &= H_k q^k + \frac{p^k p^{kT} q^k}{p^{kT} q^k} - \frac{H_k q^k q^{kT} H_k q^k}{q^{kT} H_k q^k} \\ &= H_k q^k + p^k - H_k q^k \\ &= p^k. \end{aligned}$$

We will prove (4.10) *and* (4.11) *using induction. By the last chain of equations, it follows that the assertion is true for $k = 0$. Assume that the assertion is true for*

$k-1$, i.e., assume that

$$\begin{aligned} p^{iT}Fp^j &= 0, \quad 0 \le i < j \le k-1, \\ H_k F p^j &= p^j, \quad 0 \le j \le k-1. \end{aligned} \tag{4.12}$$

We will prove the assertion for k. We have

$$\begin{aligned} f_k^\Delta &= f_{k-1}^\Delta + Fp^{k-1} \\ &= f_{k-2}^\Delta + Fp^{k-2} + Fp^{k-1} \\ &\vdots \\ &= f_{j+1}^\Delta + F(p^{j+1} + \cdots + p^{k-2} + p^{k-1}), \quad 0 \le j \le k-2. \end{aligned}$$

Then

$$\begin{aligned} p^{jT} f_k^\Delta &= p^{jT} f_{j+2}^\Delta + p^{jT} F(p^{j+1} + \cdots + p^{k-2} + p^{k-1}) \\ &= p^{jT} f_{j+1}^\Delta + p^{jT} F p^{j+1} + \cdots + p^{jT} F p^{k-2} + p^{jT} F p^{k-1}. \end{aligned}$$

Because

$$p^{jT} f_{j+1}^\Delta = 0, \quad 0 \le j \le k,$$

and by the induction assumption

$$\begin{aligned} p^{jT} F p^{j+1} &= 0, \\ &\vdots \\ p^{jT} F p^{k-2} &= 0, \\ p^{jT} F p^{k-1} &= 0, \quad 0 \le j \le k, \end{aligned}$$

we conclude that

$$p^{jT} f_k^\Delta = 0, \quad 0 \le j \le k.$$

From here, we obtain that

$$p^{jT} F H_k f_k^\Delta = 0, \quad 0 \le j < k,$$

because by the second equation of (4.12), *we find*

$$p^{jT} F H_k = p^{jT}, \quad 0 \le j < k.$$

Since

$$d^k = -H_k f_k^{\Delta}$$

and

$$\begin{aligned} p^k &= \alpha_k d^k \\ &= -\alpha_k H_k f_k^{\Delta}, \end{aligned}$$

and $\alpha_k \neq 0$, *we obtain*

$$\begin{aligned} p^{jT} F p^k &= -\alpha_k p^{jT} F H_k f_k^{\Delta} \\ &= 0, \quad 0 \le j < k, \end{aligned}$$

and from here,

$$p^{kT} F p^j = 0, \quad 0 \le j < k.$$

Next,

$$\begin{aligned} q^{kT} H_k F p^j &= q^{kT} p^j \\ &= p^{kT} F p^j \\ &= 0, \quad 0 \le j < k. \end{aligned}$$

Then

$$\begin{aligned} H_{k+1} F p^j &= \left(H_k + \frac{p^k p^{kT}}{p^{kT} q^k} - \frac{H_k q^k q^{kT} H_k}{q^{kT} H_k q^k} \right) F p^j \\ &= H_k F p^j + \frac{p^k p^{kT} F p^j}{p^{kT} q^k} - \frac{H_k q^k q^{kT} H_k F p^j}{q^{kT} H_k q^k} \\ &= H_k F p^j \\ &= p^j, \quad 0 \le j \le k. \end{aligned}$$

Thus, we have proved (4.11) *for* k. *By induction, it follows that* (4.10) *and* (4.11) *hold for any* $k \in \mathbb{N}$. *This completes the proof.*

Chapter 5

Constrained Minimization Conditions

In this chapter, we investigate minimization problems with equality and inequality constraints. They are given some first order and second order necessary and sufficiency conditions when a point is a local minimum point. Dual problems are introduced and the duality theorem is proved.

1. Constraints

Consider the general nonlinear programming problems of the form

$$\text{minimize} \quad f(x)$$

$$\text{subject to} \quad \begin{aligned} h_1(x) &= 0, \quad g_1(x) \le 0, \\ h_2(x) &= 0, \quad g_2(x) \le 0, \\ &\vdots \\ h_m(x) &= 0, \quad g_p(x) \le 0, \quad x \in \Omega, \end{aligned} \tag{5.1}$$

where $m \le n$, f, h_j, g_l, are twice continuously delta and nabla differentiable in a subset Ω of Λ^n, $j \in \{1, \ldots, m\}$, $l \in \{1, \ldots, p\}$.

For notational simplicity, we introduce the notations

$$h = (h_1, \ldots, h_m),$$

$$g = (g_1, \ldots, g_p).$$

Then, the problem (5.1) can be rewritten in the form

$$\text{minimize} \quad f(x)$$

$$\text{subject to} \quad h(x) = 0, \quad g(x) \le 0, \quad x \in \Omega.$$

Definition 15 *The constraints*

$$\begin{aligned} h(x) &= 0, \\ g(x) &\le 0, \quad x \in \Omega, \end{aligned} \tag{5.2}$$

are said to be functional constraints. The constraint $x \in \Omega$ is said to be a set constraint. A point $x \in \Omega$ that satisfies the functional constraints (5.2) *is said to be feasible.*

Definition 16 *An inequality*

$$g_l(x) \le 0, \quad l \in \{1, \ldots, p\},$$

is said to be active at a feasible point $x \in \Omega$ if

$$g_l(x) = 0$$

and inactive if

$$g_l(x) < 0.$$

A set of equality constraints on Λ^n

$$\begin{aligned} h_1(x) &= 0, \\ &\vdots \\ h_m(x) &= 0 \end{aligned} \tag{5.3}$$

defines a subset of Λ^n which is a hypersurface. If the constraints are everywhere regular, this hypersurface is of dimension $n - m$. Then, the surface defined by them is smooth.

Definition 17 *A curve on a surface S is a family of points*

$$x(t) \in S, \quad t \in [a,b], \quad a,b \in \mathbb{T}, \quad a \leq b.$$

The curve $x(t)$ is said to be Δ-differentiable (∇-differentiable) if $x^{\Delta}(t)$ ($x^{\nabla}(t)$) exists for any $t \in [a,b]$. We say that the curve pass through the point x^* if

$$x(t^*) = x^* \quad \textit{for some} \quad t^* \in [a,b].$$

Now, consider all Δ-differentiable (∇-differentiable) curves on a surface S in Λ^n passing through a point x^*.

Definition 18 *The Δ-tangent plane (∇-tangent plane) at the point x^* is defined as the collection of the Δ-derivatives (∇-derivatives) at x^* of all these Δ-differentiable (∇-differentiable) curves.*

The Δ-tangent plane (∇-tangent plane) is a subspace of Λ^n.

For surfaces defined through a set of constraint relations such as (5.3), the problem of obtaining an explicit representation for the Δ-tangent (∇-tangent) plane is a fundamental problem. Firstly, we will consider the Δ-case. Suppose that

$$\begin{aligned} &h_j(x_1(t),\ldots,x_{l-1}(t),x_l(\widetilde{\sigma}(t)),x_{l+1}(t),\ldots,x_n(t)) \\ &= h_j(x_1(t),\ldots,x_{l-1}(t),\sigma_l(x_l(t)),x_{l+1}(t),\ldots,x_n(t)), \quad t \in [a,b], \end{aligned}$$

$j \in \{1,\ldots,m\}$, $l \in \{1,\ldots,n\}$. Also, assume that h_j, $j \in \{1,\ldots,m\}$, are σ_1-completely delta differentiable. Introduce the notation

$$h^{\Delta\sigma_1\ldots\sigma_n}(x) = (h^{\Delta_1}(x), h^{\Delta_2\sigma_1}(x), h^{\Delta_3\sigma_1\sigma_2}(x), \ldots, h^{\Delta_n\sigma_1\ldots\sigma_{n-1}}(x)), \quad x \in \Lambda^n.$$

Definition 19 *A point x^* satisfying the constraint*

$$h(x^*) = 0$$

is said to be a Δ-regular point of the constraint if the vectors

$$h_1^{\Delta\sigma_1\ldots\sigma_n}(x^*), \quad \ldots, \quad h_m^{\Delta\sigma_1\ldots\sigma_n}(x^*)$$

are linearly independent.

Note that if

$$h(x) = Ax + b,$$

the Δ-regularity is equivalent to A having rank equal to m and this condition is independent of x. Define the set

$$M^\Delta = \{y \in \Lambda^n : h^{\Delta\sigma_1\ldots\sigma_n}(x^*)y = 0\}.$$

We will investigate under what conditions M^Δ is equal to the Δ-tangent plane. We have the following result.

Theorem 27 *At a Δ-regular point x^* of the surface S defined by*

$$h(x) = 0,$$

the Δ-tangent plane is equal to M^Δ.

Proof 23 *Let T be the Δ-tangent plane at x^*. We have that $T \subset M^\Delta$ whether x^* is Δ-regular or not, for any curve $x(t)$ passing through x^* at $t = t^*$ having Δ-derivative $x^\Delta(t^*)$ such that*

$$h^{\Delta\sigma_1\ldots\sigma_n}(x^*)x^\Delta(t^*) \neq 0$$

would not lie on S. Now, we will prove that $M^\Delta \subset T$. We must show that there is a curve $x(t)$ on S passing through x^* with derivative y and

$$h\left(x^* + (t - t^*)y + h^{\Delta\sigma_1\ldots\sigma_n}(x^*)^T u(t)\right) = 0. \tag{5.4}$$

Because $h^{\Delta\sigma_1\ldots\sigma_n}(x^*)$ is of full rank since x^* is Δ-regular, we have that $m \times m$ matrix

$$h^{\Delta\sigma_1\ldots\sigma_n}(x^*)h^{\Delta\sigma_1\ldots\sigma_n}(x^*)^T$$

is nonsingular. Thus, by the implicit function theorem, there is a neighbourhood of t^* for which a continuously delta differentiable function u exists. Then the curve

$$x(t) = x^* + ty + h^{\Delta\sigma_1\ldots\sigma_n}(x^*)^T u(t), \quad t \in [a,b],$$

is the needed curve that lies on S. We differentiate (5.4) *with respect to t and we get*

$$\begin{aligned} 0 &= h^{\Delta_1}\left(x^* + (t - t^*)y + h^{\Delta\sigma_1\ldots\sigma_n}(x^*)u(t)\right)\left(y_1 + h^{\Delta_1}(x^*)u_1^{\tilde{\Delta}}(t)\right) \\ &\quad + h^{\Delta_2\sigma_1}\left(x^* + (t - t^*)y + h^{\Delta\sigma_1\ldots\sigma_n}(x^*)u(t)\right)\left(y_2 + h^{\Delta_2\sigma_1}(x^*)u_2^{\tilde{\Delta}}(t)\right) \end{aligned}$$

$$
\begin{aligned}
& +\cdots \\
& +h^{\Delta_n\sigma_1\ldots\sigma_{n-1}}\left(x^*+(t-t^*)y+h^{\Delta\sigma_1\ldots\sigma_n}(x^*)u(t)\right)\left(y_n+h^{\Delta_n\sigma_1\ldots\sigma_{n-1}}(x^*)u_n^{\widetilde{\Delta}}(t)\right) \\
= \; & h^{\Delta\sigma_1\ldots\sigma_n}\left(x^*+(t-t^*)y+h^{\Delta\sigma_1\ldots\sigma_n}(x^*)u(t)\right)\left(y+h^{\Delta\sigma_1\ldots\sigma_n}(x^*)^T u^{\widetilde{\Delta}}(t)\right).
\end{aligned}
$$

In particular, when $t=t^*$, we obtain

$$
\begin{aligned}
0 &= h^{\Delta\sigma_1\ldots\sigma_n}(x^*)\left(y+h^{\Delta\sigma_1\ldots\sigma_n}(x^*)^T u^{\widetilde{\Delta}}(t^*)\right) \\
&= h^{\Delta\sigma_1\ldots\sigma_n}(x^*)y+h^{\Delta\sigma_1\ldots\sigma_n}(x^*)h^{\Delta\sigma_1\ldots\sigma_n}(x^*)^T u^{\widetilde{\Delta}}(t^*).
\end{aligned}
$$

Therefore

$$u^{\widetilde{\Delta}}(t^*)=0$$

and hence,

$$
\begin{aligned}
x^{\Delta}(t^*) &= x^*+h^{\Delta\sigma_1\ldots\sigma_n}(x^*)^T u^{\widetilde{\Delta}}(t^*) \\
&= x^*.
\end{aligned}
$$

This completes the proof.

Now, we will consider the ∇-case. Suppose that

$$
\begin{aligned}
& h_j(x_1(t),\ldots,x_{l-1}(t),x_l(\widetilde{\rho}(t)),x_{l+1}(t),\ldots,x_n(t)) \\
& = h_j(x_1(t),\ldots,x_{l-1}(t),\rho_l(x_l(t)),x_{l+1}(t),\ldots,x_n(t)), \quad t\in[a,b],
\end{aligned}
$$

$j\in\{1,\ldots,m\}$, $l\in\{1,\ldots,n\}$. Also, assume that h_j, $j\in\{1,\ldots,m\}$, are ρ_1-completely nabla differentiable. Introduce the notation

$$h^{\nabla\sigma_1\ldots\sigma_n}(x)=(h^{\nabla_1}(x),h^{\nabla_2\rho_1}(x),h^{\nabla_3\rho_1\rho_2}(x),\ldots,h^{\nabla_n\rho_1\ldots\rho_{n-1}}(x)), \quad x\in\Lambda^n.$$

Definition 20 *A point x^* satisfying the constraint*

$$h(x^*)=0$$

is said to be a ∇-regular point of the constraint if the vectors

$$h_1^{\nabla\rho_1\ldots\rho_n}(x^*), \quad \ldots, \quad h_m^{\nabla\rho_1\ldots\rho_n}(x^*)$$

are linearly independent.

Define the set

$$M^{\nabla} = \{y \in \Lambda^n : h^{\nabla \rho_1 \dots \rho_n}(x^*)y = 0\}.$$

As above, one can prove the following result.

Theorem 28 *At a ∇-regular point x^* of the surface S defined by*

$$h(x) = 0,$$

the ∇-tangent plane is equal to M^{∇}.

2. The Duality Theorem

In this section, we define the dual problem that is associated with a given problem. We define duality through the pair of problems

$$\begin{aligned} &\text{Primal} \\ &\text{minimize} \quad c^T x \\ &\text{subject to} \quad Ax \geq b, \quad x \geq 0 \end{aligned} \tag{5.5}$$

and

$$\begin{aligned} &\text{Dual} \\ &\text{maximize} \quad y^T b \\ &\text{subject to} \quad y^T A \leq c^T, \quad y \geq 0. \end{aligned} \tag{5.6}$$

If A is an $m \times m$ matrix, then x is an m dimensional column vector, b is an n dimensional column vector, c^T is an n dimensional row vector, and y^T is an m dimensional row vector. The vector x is the variable of the primal problem and y is the variable of the dual problem.

Definition 21 *The pair of problems* (5.5) *and* (5.6) *is called the symmetric form of duality.*

Note that the role of primal and dual can be reversed. Studying in detail the process for which the dual is obtained from the primal: transposition of coefficient matrix, reversal of constraint inequalities, and change of minimization or maximization, we see that this same process applied to the dual yields the primal. In other words, if the dual problem is transformed, by multiplying the objective function and the constraints by minus unity, so that it has the structure of the primal problem, its corresponding dual problem will be equivalent to the original primal problem.

Example 3 *Consider the problem*

$$\text{minimize} \quad c^T x$$

$$\text{subject to} \quad Ax = b, \quad x \geq 0.$$

We can write it in the equivalent form

$$\text{minimize} \quad c^T x$$

$$\text{subject to} \quad Ax \geq b, \quad -Ax \geq -b, \quad x \geq 0,$$

which is in the form of the primal problem (5.5) *with coefficient matrix*

$$\begin{pmatrix} A \\ -A \end{pmatrix}.$$

Using a dual vector partitioned as (u, v)*, the corresponding dual problem is*

$$\text{maximize} \quad u^T b - v^T b$$

$$\text{subject to} \quad u^T A - v^T A \leq c^T, \quad u \geq 0, \quad v \geq 0.$$

Setting

$$y = u - v,$$

we may simplify the representation of the dual problem as follows

$$\text{maximize} \quad y^T b$$

$$\text{subject to} \quad y^T A \leq c^T.$$

This is the asymmetric form of the duality relation. In this form the dual vector y *is not restricted to be nonnegative.*

Theorem 29 *(Weak Duality Theorem) If* x *and* y *are feasible for* (5.5) *and* (5.6)*, respectively, then*

$$c^T x \geq y^T b.$$

Proof 24 *We have*

$$\begin{aligned} hy^T b &\leq y^T Ax \\ &\leq c^T x. \end{aligned}$$

This completes the proof.

Corollary 1 *If x^0 and y^0 are feasible for (5.5) and (5.6), respectively, and if*

$$c^T x^0 = y^{0T} b,$$

then x^0 and y^0 are optimal for their respectve problems.

Theorem 30 *(The Duality Theorem) If either the problems (5.5) or (5.6) has a finite optimal solution, so does the either, and the corresponding values of the objective functions are equal. If either problem has an unbounded objective function, then other problem has no feasible solution.*

Proof 25 *Suppose that the primal problem (5.5) has a finite optimal solution with value z_0. In $\mathbb{R}^{m+1}$, define*

$$C = \{(r,w) : t = tz_0 - c^T x, \quad w = tb - Ax, \quad x \geq 0, \quad t \geq 0\}.$$

We will prove that $(1,0) \notin C$. If

$$\begin{aligned} w &= t_0 b - Ax_0 \\ &= 0 \quad \text{with} \quad t_0 > 0, \quad x_0 \geq 0, \end{aligned}$$

then $\frac{x_0}{t_0}$ is feasible for the primal problem (5.5) and hence,

$$\begin{aligned} \frac{r}{t_0} &= z_0 - c^T \frac{x_0}{t_0} \\ &\leq 0, \end{aligned}$$

we conclude that $r \leq 0$. If

$$\begin{aligned} w &= -Ax_0 \\ &= 0 \quad \text{with} \quad x_0 \geq 0, \quad c^T x_0 = -1, \end{aligned}$$

and if x is any feasible solution to the primal problem (5.5), then $x+\alpha x_0$ is feasible for any $\alpha \geq 0$ and gives arbitrarily small objective values as α increases. This contradicts with our assumption on the existence of a finite optimal solution and thus, there is no such x_0. Therefore $(1,0) \notin C$. Then there is a hyperplane separating $(1,0)$ and C. Hence, there is a nonzero vector $[s,y] \in \mathbb{R}^{m+1}$ and a constant c such that

$$\begin{aligned} s &< c \\ &= \inf\{sr+y^T w : (r,w) \in C\}. \end{aligned}$$

Since C is a closed cone, we conclude that $c \geq 0$. If there were $(r,w) \in C$ such that

$$sr+y^T w < 0,$$

then $\alpha(r,w)$ for large α would violate the hyperplane inequality. Since $(0,0) \in C$, we must have $c \leq 0$. Thus, $c=0$ and $s<0$. Without loss of generality, assume that $c=-1$. Therefore

$$-r+y^T w \geq 0$$

for any $(r,w) \in C$. Equivalently, using the definition of C,

$$-tz_0+c^T x+ty^T b-y^T Ax \geq 0,$$

or

$$(c^T - y^T A)x - tz_0 + ty^T b \geq 0$$

for any $x \geq 0$ and $t \geq 0$. Setting $t=0$ yields

$$y^T Ax \leq c^T$$

and then y is feasible for the dual problem (5.6). Setting $x=0$ and $t=1$ yields

$$y^T b \geq z_0$$

and then y is optimal for the dual problem (5.6).

If the primal problem is unbounded and y is feasible for the dual problem (5.6) we must have

$$y^T b \leq -M$$

for arbitrarily large M, which is impossible. This completes the proof.

3. Equality Constraints

We begin by deriving the first order necessary conditions for a point to be a local extremum subject to equality constraints.

Theorem 31 *Let x^* be a Δ-regular and ∇-regular point of the constraints*

$$h(x) = 0$$

and a local minimum point of the function f subject to these constraints. Then, for any $y, z \in \Lambda^n$ satisfying

$$h^{\Delta\sigma_1\ldots\sigma_n}(x^*)y = 0$$

and

$$h^{\nabla\rho_1\ldots\rho_n}(x^*)z = 0,$$

must also satisfy

$$f^{\Delta\sigma_1\ldots\sigma_n}(x^*)y \geq 0$$

and

$$f^{\nabla\rho_1\ldots\rho_n}(x^*)z \leq 0.$$

Proof 26 *Let y be a vector in the Δ-tangent plane at x^* and z be a vector in the ∇-tangent plane at x^*. Suppose that $x(t)$ is a any Δ-smooth and ∇-smooth curve on the constraint surface passing through x^* with Δ-derivative y at x^* and ∇-derivative z at x^*, i.e., we have*

$$\begin{aligned} x(t^*) &= x^*, \\ x^{\Delta}(t^*) &= y, \\ x^{\nabla}(t^*) &= z \end{aligned}$$

and

$$h(x(t)) = 0, \quad t \in [a, b].$$

Since x^* is a Δ-regular point and a ∇-regular point, the Δ-tangent plane is identical with the set of y's for which

$$h^{\Delta\sigma_1\ldots\sigma_n}(x^*)y = 0$$

and the ∇-tangent plane is equal to the set of z's for which

$$h^{\nabla\rho_1\ldots\rho_n}(x^*)z = 0.$$

Then, since x^* is a constrained local minimum of f, we have

$$\frac{\Delta}{\Delta t} f(x(t))\bigg|_{t=t^*} \geq 0$$

and

$$\frac{\nabla}{\nabla t} f(x(t))\bigg|_{t=t^*} \leq 0,$$

or equivalently

$$f^{\Delta\sigma_1\ldots\sigma_n}(x(t^*))x^{\Delta}(t^*) \geq 0$$

and

$$f^{\nabla\rho_1\ldots\rho_n}(x(t^*))x^{\nabla}(t^*) \leq 0,$$

or equivalently

$$f^{\Delta\sigma_1\ldots\sigma_n}(x^*)y \geq 0$$

and

$$f^{\nabla\rho_1\ldots\rho_n}(x^*)z \leq 0.$$

This completes the proof.

As above, one can prove the following result regarding a local maximum point.

Theorem 32 *Let x^* be a Δ-regular and ∇-regular point of the constraints*

$$h(x) = 0$$

and a local maximum point of the function f subject to these constraints. Then, for any $y, z \in \Lambda^n$ satisfying

$$h^{\Delta\sigma_1\ldots\sigma_n}(x^*)y = 0$$

and

$$h^{\nabla\rho_1\ldots\rho_n}(x^*)z = 0,$$

must also satisfy

$$f^{\Delta\sigma_1\ldots\sigma_n}(x^*)y \leq 0$$

and

$$f^{\nabla\rho_1\ldots\rho_n}(x^*)z \geq 0.$$

Theorem 33 *Let x^* be a local minimum point for the objective function f subject to the constraints*

$$h(x) = 0.$$

Assume that x^* is a Δ-regular and ∇-regular point of these constraints. Then there is a $\lambda \in \mathbb{R}^m$ such that

$$f^{\Delta\sigma_1\ldots\sigma_n}(x^*) + \lambda^T h^{\Delta\sigma_1\ldots\sigma_n}(x^*) \geq 0 \tag{5.7}$$

and

$$f^{\nabla\rho_1\ldots\rho_n}(x^*) + \lambda^T h^{\nabla\rho_1\ldots\rho_n}(x^*) \leq 0. \tag{5.8}$$

Proof 27 *By Theorem 31, it follows that the values of the linear problems*

$$\text{minimize} \quad f^{\Delta\sigma_1\ldots\sigma_n}(x^*)y$$

$$\text{subject to} \quad h^{\Delta\sigma_1\ldots\sigma_n}(x^*)y = 0$$

and

$$\text{maximize} \quad f^{\nabla\rho_1\ldots\rho_n}(x^*)y$$

$$\text{subject to} \quad h^{\nabla\rho_1\ldots\rho_n}(x^*)y = 0$$

are zero. Then, by the duality theorem, Theorem 30, the dual problems are feasible, i.e., there is a $\lambda \in \mathbb{R}^m$ such that (5.7) and (5.8) hold. This completes the proof.

As above, one can prove the following result.

Theorem 34 *Let x^* be a local maximum point for the objective function f subject to the constraints*

$$h(x) = 0.$$

Assume that x^* is a Δ-regular and ∇-regular point of these constraints. Then there is a $\lambda \in \mathbb{R}^m$ such that

$$f^{\Delta\sigma_1\ldots\sigma_n}(x^*) + \lambda^T h^{\Delta\sigma_1\ldots\sigma_n}(x^*) \leq 0 \tag{5.9}$$

and

$$f^{\nabla\rho_1\ldots\rho_n}(x^*) + \lambda^T h^{\nabla\rho_1\ldots\rho_n}(x^*) \geq 0. \tag{5.10}$$

It is convenient to introduce the Lagrangian associated with the constrained problem, defined by

$$l(x,\lambda) = f(x) + \lambda^T h(x).$$

Then, the conditions (5.7) and (5.8) can be rewritten as follows

$$\begin{aligned} l_x^{\Delta\sigma_1\ldots\sigma_n}(x,\lambda) &\geq 0 \\ l_x^{\nabla\rho_1\ldots\rho_n}(x,\lambda) &\leq 0 \\ l'_\lambda(x,\lambda) &= 0. \end{aligned} \tag{5.11}$$

The conditions (5.9) and (5.10) can be rewritten as follows

$$\begin{aligned} l_x^{\Delta\sigma_1\ldots\sigma_n}(x,\lambda) &\leq 0 \\ l_x^{\nabla\rho_1\ldots\rho_n}(x,\lambda) &\geq 0 \\ l'_\lambda(x,\lambda) &= 0. \end{aligned} \tag{5.12}$$

Definition 22 *The components of the vector $\lambda \in \mathbb{R}^m$ for which* (5.11) *or* (5.12) *hold are said to be Lagrange multipliers.*

Example 4 *Let $\mathbb{T}_1 = \mathbb{Z}$, $\mathbb{T}_2 = 2^{\mathbb{N}_0}$, $\mathbb{T}_3 = 3^{\mathbb{N}_0}$. Consider the problem*

$$\textit{minimize} \quad x_1x_2 + x_2x_3 + x_1x_3$$

$$\textit{subject to} \quad x_1 + x_2 + x_3 = 3.$$

Here

$$\begin{aligned} \sigma_1(x_1) &= x_1 + 1, \quad x_1 \in \mathbb{T}_1, \\ \sigma_2(x_2) &= 2x_2, \quad x_2 \in \mathbb{T}_2, \\ \sigma_3(x_3) &= 3x_3, \quad x_3 \in \mathbb{T}_3, \\ \rho_1(x_1) &= x_1 - 1, \\ \rho_2(x_2) &= \frac{1}{2}x_2, \quad x_2 \in \mathbb{T}_2, \quad x_2 \neq 1, \\ \rho_2(1) &= 1, \end{aligned}$$

$$\rho_3(x_3) = \frac{1}{3}x_3, \quad x_3 \in \mathbb{T}_3, \quad x_3 \neq 1,$$

$$\rho_3(1) = 1$$

and

$$l(x_1,x_2,x_3,\lambda) = x_1x_2 + x_2x_3 + x_1x_3 + \lambda(x_1 + x_2 + x_3),$$

$(x_1,x_2,x_3) \in \Lambda^3$, $\lambda \in \mathbb{R}$. *Then, we have the following*

$$\begin{aligned}
l_{x_1}^{\Delta_1}(x_1,x_2,x_3,\lambda) &= x_2 + x_3 + \lambda, \\
l_{x_2}^{\Delta_2\sigma_1}(x_1,x_2,x_3,\lambda) &= \sigma_1(x_1) + x_3 + \lambda \\
&= x_1 + 1 + x_3 + \lambda \\
&= x_1 + x_3 + \lambda + 1, \\
l_{x_3}^{\Delta_3\sigma_1\sigma_2}(x_1,x_2,x_3,\lambda) &= \sigma_2(x_2) + \sigma_1(x_1) + \lambda \\
&= x_1 + 1 + 2x_2 + \lambda \\
&= x_1 + 2x_2 + \lambda + 1, \quad (x_1,x_2,x_3,\lambda) \in \Lambda^3 \times \mathbb{R},
\end{aligned}$$

and

$$\begin{aligned}
l_{x_1}^{\nabla_1}(x_1,x_2,x_3,\lambda) &= x_2 + x_3 + \lambda, \\
l_{x_2}^{\nabla_2\rho_1}(x_1,x_2,x_3,\lambda) &= \rho_1(x_1) + x_3 + \lambda \\
&= x_1 - 1 + x_3 + \lambda \\
&= x_1 + x_3 + \lambda - 1, \\
l_{x_3}^{\nabla_3\rho_1\rho_2}(x_1,x_2,x_3,\lambda) &= \rho_2(x_2) + \rho_1(x_1) + \lambda \\
&= \begin{cases} x_1 - 1 + \frac{1}{2}x_2 + \lambda & \text{if} \quad x_2 \neq 1, \quad x_2 \in \mathbb{T}_2, \\ x_1 - 1 + 1 + \lambda & \text{if} \quad x_2 = 1 \end{cases}
\end{aligned}$$

$$= \begin{cases} x_1 + \frac{1}{2}x_2 + \lambda - 1 & \textit{if} \quad x_2 \neq 1, \quad x_2 \in \mathbb{T}_2, \\ x_1 + \lambda \quad \textit{if} \quad x_2 = 1. \end{cases}$$

Consider the following cases.

1. *Let* $x_2 \in \mathbb{T}_2$, $x_2 \neq 1$. *Then, the system* (5.11) *takes the form*

$$\begin{aligned} x_2 + x_3 + \lambda &\geq 0 \\ x_1 + x_3 + \lambda + 1 &\geq 0 \\ x_1 + 2x_2 + \lambda + 1 &\geq 0 \\ x_2 + x_3 + \lambda &\leq 0 \\ x_1 + x_3 + \lambda - 1 &\leq 0 \\ x_1 + \frac{1}{2}x_2 + \lambda - 1 &\leq 0. \end{aligned}$$

We have that

$$\begin{aligned} x_1 &= 1 \\ x_2 &= 1 \\ x_3 &= 1 \\ \lambda &= -2, \end{aligned}$$

which is not a solution because $x_2 \neq 1$.

2. *Let* $x_2 = 1$. *Then* (5.11) *takes the form*

$$\begin{aligned} x_3 + \lambda + 1 &\geq 0 \\ x_1 + x_3 + \lambda + 1 &\geq 0 \end{aligned}$$

$$\begin{aligned} x_1+\lambda+3 &\geq 0 \\ x_3+\lambda+1 &\leq 0 \\ x_1+x_3+\lambda-1 &\leq 0 \\ x_1+\lambda &\leq 0. \end{aligned}$$

Then

$$\begin{aligned} x_1 &= 1 \\ x_2 &= 1 \\ x_3 &= 1 \\ \lambda &= -2 \end{aligned}$$

is a solution of the considered problem.

4. Second Order Conditions

In this section, suppose that f and h are twice continuously delta differentiable and σ_j, $j \in \{1,\ldots,n\}$, are Δ_j-differentiable. In addition, assume that the function $x(t)$ is twice continuously $\widetilde{\Delta}$-differentiable and if

$$x(t) = (x_1(t),\ldots,x_n(t)),$$

then

$$x_j(\widetilde{\sigma}(t)) = \sigma_j(x_j(t)), \quad j \in \{1,\ldots,n\}.$$

Then, we have the following

$$\begin{aligned} f^{\widetilde{\Delta}}(x(t)) &= f_{x_1}^{\Delta_1}(x(t))x_1^{\widetilde{\Delta}}(t) + f_{x_2}^{\Delta_2\sigma_1}(x(t))x_2^{\widetilde{\Delta}}(t) \\ &\quad + \cdots \end{aligned}$$

$$
\begin{aligned}
&+f_{x_n}^{\Delta_n\sigma_1\ldots\sigma_{n-1}}(x(t))x_n^{\widetilde{\Delta}}(t)\\
&= f^{\Delta\sigma_1\ldots\sigma_n}(x(t))x^{\widetilde{\Delta}}(t)
\end{aligned}
$$

and

$$
\begin{aligned}
f^{\widetilde{\Delta}^2}(x(t)) &= \left(f^{\Delta\sigma_1\ldots\sigma_n}(x(t))\right)^{\sigma_1\ldots\sigma_n} x^{\widetilde{\Delta}^2}(t)\\
&+\Bigg(f_{x_1x_1}^{\Delta_1^2}(x(t))x_1^{\widetilde{\Delta}}(t)+f_{x_1x_2}^{\Delta_1\Delta_2\sigma_1}(x(t))x_2^{\widetilde{\Delta}}(t)\\
&+\cdots+f_{x_1x_2}^{\Delta_1\Delta_n\sigma_1\ldots\sigma_{n-1}}(x(t))x_n^{\widetilde{\Delta}}(t)\Bigg)x_1^{\widetilde{\Delta}}(t)\\
&+\Bigg(f_{x_2x_1}^{\Delta_2\sigma_1\Delta_1}(x(t))x_1^{\widetilde{\Delta}}(t)+f_{x_2x_2}^{\Delta_2\sigma_1\Delta_2\sigma_1}(x(t))x_2^{\widetilde{\Delta}}(t)\\
&+\cdots+f_{x_2x_2}^{\Delta_1\sigma_1\Delta_n\sigma_1\ldots\sigma_{n-1}}(x(t))x_n^{\widetilde{\Delta}}(t)\Bigg)x_2^{\widetilde{\Delta}}(t)\\
&+\cdots\\
&+\Bigg(f_{x_nx_1}^{\Delta_n\sigma_1\ldots\sigma_{n-1}\Delta_1}(x(t))x_1^{\widetilde{\Delta}}(t)+f_{x_nx_2}^{\Delta_n\sigma_1\ldots\sigma_{n-1}\Delta_2\sigma_1}(x(t))x_2^{\widetilde{\Delta}}(t)\\
&+\cdots+f_{x_nx_n}^{\Delta_n\sigma_1\ldots\sigma_{n-1}\Delta_n\sigma_1\ldots\sigma_{n-1}}(x(t))x_n^{\widetilde{\Delta}}(t)\Bigg)x_n^{\widetilde{\Delta}}(t).
\end{aligned}
$$

Define the matrices

$$
F_1(x(t))=\left(f^{\Delta\sigma_1\ldots\sigma_n}(x(t))\right)^{\sigma_1\ldots\sigma_n}
$$

and

$$
F_2(x(t))=\begin{pmatrix}
f_{x_1x_1}^{\Delta_1^2}(x(t)) & f_{x_1x_2}^{\Delta_1\Delta_2\sigma_1}(x(t)) & \cdots & f_{x_1x_n}^{\Delta_1\Delta_n\sigma_1\ldots\sigma_n}(x(t))\\
f_{x_2x_1}^{\Delta_2\sigma_1\Delta_1}(x(t)) & f_{x_2x_2}^{\Delta_2\sigma_1\Delta_2\sigma_1}(x(t)) & \cdots & f_{x_2x_n}^{\Delta_2\sigma_1\Delta_n\sigma_1\ldots\sigma_{n-1}}(x(t))\\
\vdots & \vdots & \vdots & \vdots\\
f_{x_nx_1}^{\Delta_n\sigma_1\ldots\sigma_{n-1}\Delta_1}(x(t)) & f_{x_nx_2}^{\Delta_n\sigma_1\ldots\sigma_{n-1}\Delta_2\sigma_1}(x(t)) & \cdots & f_{x_nx_n}^{\Delta_n\sigma_1\ldots\sigma_{n-1}\Delta_n\sigma_1\ldots\sigma_{n-1}}(x(t))
\end{pmatrix}.
$$

Then

$$
f^{\widetilde{\Delta}^2}(t)=F_1(x(t))x^{\widetilde{\Delta}^2}(t)+x^{\widetilde{\Delta}}(t)^T F_2(x(t))x^{\widetilde{\Delta}}(t).
$$

As above, if

$$
H_1(x(t))=\left(h^{\Delta\sigma_1\ldots\sigma_n}(x(t))\right)^{\sigma_1\ldots\sigma_n}
$$

and

$$H_{2j}(x(t)) = \begin{pmatrix} h^{\Delta_1^2}_{jx_1x_1}(x(t)) & h^{\Delta_1\Delta_2\sigma_1}_{jx_1x_2}(x(t)) & \cdots & h^{\Delta_1\Delta_n\sigma_1\ldots\sigma_n}_{jx_1x_n}(x(t)) \\ h^{\Delta_2\sigma_1\Delta_1}_{jx_2x_1}(x(t)) & h^{\Delta_2\sigma_1\Delta_2\sigma_1}_{jx_2x_2}(x(t)) & \cdots & h^{\Delta_2\sigma_1\Delta_n\sigma_1\ldots\sigma_{n-1}}_{jx_2x_n}(x(t)) \\ \vdots & \vdots & \vdots & \vdots \\ h^{\Delta_n\sigma_1\ldots\sigma_{n-1}\Delta_1}_{jx_nx_1}(x(t)) & h^{\Delta_n\sigma_1\ldots\sigma_{n-1}\Delta_2\sigma_1}_{jx_nx_2}(x(t)) & \cdots & h^{\Delta_n\sigma_1\ldots\sigma_{n-1}\Delta_n\sigma_1\ldots\sigma_{n-1}}_{jx_nx_n}(x(t)) \end{pmatrix},$$

$j \in \{1,\ldots,m\}$, and

$$H_2(x(t)) = \begin{pmatrix} H_{21}(x(t)) \\ H_{22}(x(t)) \\ \vdots \\ H_{2m}(x(t)) \end{pmatrix},$$

then

$$h^{\widetilde{\Delta}^2}(t) = H_1(x(t))x^{\widetilde{\Delta}^2} + x^{\widetilde{\Delta}}(t)^T H_2(x(t))x^{\widetilde{\Delta}}(t).$$

Moreover, if

$$F_1^1(x(t)) = \left(f^{\nabla\rho_1\ldots\rho_n}(x(t))\right)^{\rho_1\ldots\rho_n}$$

and

$$F_2^1(x(t)) = \begin{pmatrix} f^{\nabla_1^2}_{x_1x_1}(x(t)) & f^{\nabla_1\nabla_2\rho_1}_{x_1x_2}(x(t)) & \cdots & f^{\nabla_1\nabla_n\rho_1\ldots\rho_n}_{x_1x_n}(x(t)) \\ f^{\nabla_2\rho_1\nabla_1}_{x_2x_1}(x(t)) & f^{\nabla_2\rho_1\nabla_2\rho_1}_{x_2x_2}(x(t)) & \cdots & f^{\nabla_2\rho_1\nabla_n\rho_1\ldots\rho_{n-1}}_{x_2x_n}(x(t)) \\ \vdots & \vdots & \vdots & \vdots \\ f^{\nabla_n\rho_1\ldots\rho_{n-1}\nabla_1}_{x_nx_1}(x(t)) & f^{\nabla_n\rho_1\ldots\rho_{n-1}\nabla_2\rho_1}_{x_nx_2}(x(t)) & \cdots & f^{\nabla_n\rho_1\ldots\rho_{n-1}\nabla_n\rho_1\ldots\rho_{n-1}}_{x_nx_n}(x(t)) \end{pmatrix},$$

and

$$H_1^1(x(t)) = \left(h^{\nabla\rho_1\ldots\rho_n}(x(t))\right)^{\rho_1\ldots\rho_n},$$

and

$$H_{2j}^1(x(t)) = \begin{pmatrix} h^{\nabla_1^2}_{jx_1x_1}(x(t)) & h^{\nabla_1\nabla_2\rho_1}_{jx_1x_2}(x(t)) & \cdots & h^{\nabla_1\nabla_n\rho_1\ldots\rho_n}_{jx_1x_n}(x(t)) \\ h^{\nabla_2\rho_1\nabla_1}_{jx_2x_1}(x(t)) & h^{\nabla_2\rho_1\nabla_2\rho_1}_{jx_2x_2}(x(t)) & \cdots & h^{\nabla_2\rho_1\nabla_n\rho_1\ldots\rho_{n-1}}_{jx_2x_n}(x(t)) \\ \vdots & \vdots & \vdots & \vdots \\ h^{\nabla_n\rho_1\ldots\rho_{n-1}\nabla_1}_{jx_nx_1}(x(t)) & h^{\nabla_n\rho_1\ldots\rho_{n-1}\nabla_2\rho_1}_{jx_nx_2}(x(t)) & \cdots & h^{\nabla_n\rho_1\ldots\rho_{n-1}\nabla_n\rho_1\ldots\rho_{n-1}}_{jx_nx_n}(x(t)) \end{pmatrix},$$

$j \in \{1,\ldots,m\}$, and

$$H_2^1(x(t) = \begin{pmatrix} H_{21}^1(x(t)) \\ H_{22}^1(x(t)) \\ \vdots \\ H_{2m}^1(x(t)) \end{pmatrix},$$

then

$$f^{\widetilde{\nabla}^2}(t) = F_1^1(x(t))x^{\widetilde{\nabla}^2}(t) + x^{\widetilde{\nabla}}(t)^T F_2^1(x(t))x^{\widetilde{\nabla}}(t)$$

and

$$h^{\widetilde{\nabla}^2}(t) = H_1^1(x(t))x^{\widetilde{\nabla}^2}(t) + x^{\widetilde{\nabla}}(t)^T H_2^1(x(t))x^{\widetilde{\nabla}}(t).$$

Theorem 35 *(Second Order Necessary Conditions) Suppose that x^* is a local minimum point of the objective function f subject to*

$$h(x) = 0$$

and x^* is a Δ-regular and ∇-regular point for these constraints. Then, there is a $\lambda \in \mathbb{R}^m$ such that

$$F_1(x^*) + \lambda^T H_1(x^*)x^{\widetilde{\Delta}^2}(t^*) = 0. \tag{5.13}$$

If we denote by M the Δ-tangent plane

$$M = \{y \in \Lambda^m : h^{\Delta\sigma_1\ldots\sigma_n}(x^*)y = 0\},$$

then the matrix

$$L(x^*) = F_2(x^*) + \lambda^T H_2(x^*)$$

is positive semidefinite on M, i.e.,

$$y^T L(x^*)y \geq 0 \quad \textit{for any} \quad y \in M.$$

Proof 28 *The value of the linear problem*

$$\textit{minimize} \quad F_1(x^*)y$$

$$\textit{subject to} \quad H_1(x^*)y = 0$$

is zero. Then, applying the duality theorem, Theorem 30, the dual problem is feasible. Therefore there is a $\lambda \in \mathbb{R}^m$ such that (5.13) holds. Note that

$$f^{\widetilde{\Delta}^2}(x^*) \geq 0.$$

By definition, we find

$$f^{\widetilde{\Delta}^2}(x^*) = F_1(x^*)x^{\widetilde{\Delta}^2}(t^*) + x^{\widetilde{\Delta}}(t^*)^T F_2(x^*)x^{\widetilde{\Delta}}(t^*). \tag{5.14}$$

If we $\widetilde{\Delta}$-differentiate the relation

$$\lambda^T h(x(t)) = 0$$

twice, we obtain

$$\lambda^T H_1(x^*)x^{\widetilde{\Delta}^2}(t^*) + x^{\widetilde{\Delta}}(t^*)^T \lambda^T H_2(x^*)x^{\widetilde{\Delta}}(t^*) = 0. \tag{5.15}$$

Adding (5.14) *and* (5.15), *we arrive at*

$$
\begin{aligned}
0 &\leq F_1(x^*)x^{\widetilde{\Delta}^2}(t^*) + x^{\widetilde{\Delta}}(t^*)^T F_2(x^*)x^{\widetilde{\Delta}}(t^*)\\
&\quad + \lambda^T H_1(x^*)x^{\widetilde{\Delta}^2}(t^*) + x^{\widetilde{\Delta}}(t^*)^T \lambda^T H_2(x^*)x^{\widetilde{\Delta}}(t^*)\\
&= \left(F_1(x^*) + \lambda^T H_1(x^*)\right) x^{\widetilde{\Delta}^2}(t^*)\\
&\quad + x^{\widetilde{\Delta}}(t^*)^T \left(F_2(x^*) + \lambda^T H_2(x^*)\right) x^{\widetilde{\Delta}}(t^*)\\
&= x^{\widetilde{\Delta}}(t^*)^T L(x^*) x^{\widetilde{\Delta}}(t^*)\\
&\geq 0.
\end{aligned}
$$

Because $x^{\widetilde{\Delta}}(t^*)$ *was arbitrarily chosen in* M, *the proof is completed.*

As above, one can prove the following results.

Theorem 36 *(Second Order Necessary Conditions) Suppose that* x^* *is a local minimum point of the objective function* f *subject to*

$$h(x) = 0$$

and x^* *is a* Δ*-regular and* ∇*-regular point for these constraints. Then, there is a* $\lambda \in \mathbb{R}^m$ *such that*

$$F_1^1(x^*) + \lambda^T H_1^1(x^*)x^{\widetilde{\nabla}^2}(t^*) = 0.$$

If we denote by M^1 *the* ∇*-tangent plane*

$$M^1 = \{y \in \Lambda^m : h^{\nabla \rho_1 \ldots \rho_n}(x^*)y = 0\},$$

then the matrix

$$L^1(x^*) = F_2^1(x^*) + \lambda^T H_2^1(x^*)$$

is positive semidefinite on M^1, *i.e.,*

$$y^T L^1(x^*)y \geq 0 \quad \text{for any} \quad y \in M^1.$$

Theorem 37 *(Second Order Necessary Conditions) Suppose that* x^* *is a local maximum point of the objective function* f *subject to*

$$h(x) = 0$$

and x^* is a Δ-regular and ∇-regular point for these constraints. Then, there is a $\lambda \in \mathbb{R}^m$ such that

$$F_1(x^*) + \lambda^T H_1(x^*) x^{\widetilde{\nabla}^2}(t^*) = 0.$$

If we denote by M^1 the Δ-tangent plane

$$M = \{y \in \Lambda^m : h^{\Delta\sigma_1 \ldots \sigma_n}(x^*) y = 0\},$$

then the matrix

$$L(x^*) = F_2(x^*) + \lambda^T H_2(x^*)$$

is negative semidefinite on M, i.e.,

$$y^T L(x^*) y \geq 0 \quad \textit{for any} \quad y \in M.$$

Theorem 38 *(Second Order Necessary Conditions) Suppose that x^* is a local maximum point of the objective function f subject to*

$$h(x) = 0$$

and x^* is a Δ-regular and ∇-regular point for these constraints. Then, there is a $\lambda \in \mathbb{R}^m$ such that

$$F_1^1(x^*) + \lambda^T H_1^1(x^*) x^{\widetilde{\nabla}^2}(t^*) = 0.$$

If we denote by M^1 the ∇-tangent plane

$$M^1 = \{y \in \Lambda^m : h^{\nabla\rho_1 \ldots \rho_n}(x^*) y = 0\},$$

then the matrix

$$L^1(x^*) = F_2^1(x^*) + \lambda^T H_2^1(x^*)$$

is negative semidefinite on M^1, i.e.,

$$y^T L^1(x^*) y \leq 0 \quad \textit{for any} \quad y \in M^1.$$

Consider the function $f(x(t))$ as a function of $t \in [a,b]$. Then, applying the Taylor formula, we find

$$\begin{aligned} f(x(t)) &= f(x^*) + f^{\Delta\sigma_1 \ldots \sigma_n}(x^*)(t - t^*) \\ &+ \int_{t^*}^{t} \Big(F_1(x(\tau)) x^{\widetilde{\Delta}^2}(\tau) + x^{\widetilde{\Delta}}(\tau)^T F_2(x(\tau)) x^{\widetilde{\Delta}}(\tau) \Big) h_1(t, \widetilde{\sigma}(\tau)) \widetilde{\Delta}\tau \end{aligned}$$

and

$$\begin{aligned} f(x(t)) &= f(x^*)+f^{\nabla\rho_1\ldots\rho_n}(x^*)(t-t^*) \\ &\quad + \int_{t^*}^{t} \left(F_1^1(x(\tau))x^{\widetilde{\nabla}^2}(\tau)+x^{\widetilde{\nabla}}(\tau)^T F_2^1(x(\tau))x^{\widetilde{\nabla}}(\tau)\right) h_1(t,\widetilde{\nabla}(\tau))\widetilde{\nabla}\tau. \end{aligned}$$

Next, we will deduct some sufficiency conditions for a local extremum point.

Theorem 39 *(Second Order Sufficiency Conditions) Suppose that there is a point x^* such that*

$$h(x^*) = 0$$

and a $\lambda \in \mathbb{R}^m$ such that

$$\begin{aligned} f^{\Delta\sigma_1\ldots\sigma_n}(x^*)+\lambda^T h^{\Delta\sigma_1\ldots\sigma_n}(x^*) &\geq 0, \\ f^{\nabla\rho_1\ldots\rho_n}(x^*)+\lambda^T h^{\nabla\rho_1\ldots\rho_n}(x^*) &\leq 0, \\ F_1(x^*)+\lambda^T H_1(x^*) &\geq 0, \\ F_1^1(x^*)+\lambda^T H_1^1(x^*) &\leq 0. \end{aligned}$$

Suppose also, that

$$L(x^*) = F_2(x^*)+\lambda^T H_2(x^*)$$

and

$$L^1(x^*) = F_2^1(x^*)+\lambda^T H_2^1(x^*)$$

are positive definite on

$$M = \{y \in \Lambda^m : h^{\Delta\sigma_1\ldots\sigma_n}(x^*)y = 0\}$$

and

$$M^1 = \{y \in \Lambda^m : h^{\nabla\rho_1\ldots\rho_n}(x^*)y = 0\},$$

respectively, that is

$$y^T L(x^*)y > 0$$

and

$$z^T L^1(x^*)z > 0,$$

respectively, for any $y \in M$, $y \neq 0$ and $z \in M^1$, $z \neq 0$. Then x^* is a local minimum point for the objective function f subject to

$$h(x) = 0.$$

Proof 29 *Assume the contrary, i.e., assume that x^* is not a local minimum point of f.*

1. *Let x^* be right-scattered and $\sigma(x^*)$ is feasible, and*
$$f(\sigma(x^*)) < f(x^*).$$
Let also, $\sigma(x^*) = x(t_1)$ for some $t_1 > t^*$. We have
$$\begin{aligned} h_j(\sigma(x^*)) &= 0,\\ h_j(x^*) &= 0, \quad j \in \{1,\ldots,m\}. \end{aligned}$$
Then
$$\begin{aligned} 0 &= h_j(\sigma(x^*)) - h_j(x^*)\\ &= h_j^{\Delta\sigma_1\ldots\sigma_n}(x^*)(\sigma(x^*) - x^*), \quad j \in \{1,\ldots,m\}. \end{aligned}$$
On the other hand, by the Taylor formula, we get
$$\begin{aligned} 0 &= h_j(\sigma(x^*)) - h_j(x^*)\\ &= h_j^{\Delta\sigma_1\ldots\sigma_n}(x^*)(\widetilde{\sigma}(t^*) - t^*)\\ &\quad + \int_{t^*}^{\widetilde{\sigma}(t^*)} \Big(H_{1j}(x(\tau))x^{\widetilde{\Delta}^2}(\tau) + x^{\widetilde{\Delta}}(\tau)^T H_{2j}(x(\tau))x^{\widetilde{\Delta}}(\tau)\Big) h_1(\widetilde{\sigma}(t^*), \sigma(\tau))\widetilde{\Delta}\tau, \end{aligned}$$
$j \in \{1,\ldots,m\}$. We multiply by λ_j, $j \in \{1,\ldots,m\}$, and we find
$$\begin{aligned} 0 &= \lambda_j h_j^{\Delta\sigma_1\ldots\sigma_n}(x^*)(\widetilde{\sigma}(t^*) - t^*)\\ &\quad + \int_{t^*}^{\widetilde{\sigma}(t^*)} \Big(\lambda_j H_{1j}(x(\tau))x^{\widetilde{\Delta}^2}(\tau) + \lambda_j x^{\widetilde{\Delta}}(\tau)^T H_{2j}(x(\tau))x^{\widetilde{\Delta}}(\tau)\Big) h_1(\widetilde{\sigma}(t^*), \sigma(\tau))\widetilde{\Delta}\tau, \end{aligned}$$
$j \in \{1,\ldots,m\}$. Therefore
$$\begin{aligned} 0 &= \lambda^T h^{\Delta\sigma_1\ldots\sigma_n}(x^*)(\widetilde{\sigma}(t^*) - t^*)\\ &\quad + \int_{t^*}^{\widetilde{\sigma}(t^*)} \Big(\lambda^T H_1(x(\tau))x^{\widetilde{\Delta}^2}(\tau) + x^{\widetilde{\Delta}}(\tau)^T(\lambda^T H_2(x(\tau)))x^{\widetilde{\Delta}}(\tau)\Big) h_1(\widetilde{\sigma}(t^*), \sigma(\tau))\widetilde{\Delta}\tau. \end{aligned} \tag{5.16}$$
Now, we apply the Taylor formula for f and we find
$$0 > f(\sigma(x^*)) - f(x^*)$$

$$
\begin{aligned}
&= f^{\Delta\sigma_1\ldots\sigma_n}(x^*)(\widetilde{\sigma}(t^*)-t^*) \\
&\quad + \int_{t^*}^{\widetilde{\sigma}(t^*)} \left(F_1(x(\tau))x^{\widetilde{\Delta}^2}(\tau) + x^{\widetilde{\Delta}}(\tau)^T F_2(x(\tau))x^{\widetilde{\Delta}}(\tau)\right) h_1(\widetilde{\sigma}(t^*),\sigma(\tau))\widetilde{\Delta}\tau.
\end{aligned}
$$

We add the last inequality and (5.16) and we arrive at

$$
\begin{aligned}
0 &> \lambda^T h^{\Delta\sigma_1\ldots\sigma_n}(x^*)(\widetilde{\sigma}(t^*)-t^*) \\
&\quad + \int_{t^*}^{\widetilde{\sigma}(t^*)} \left(\lambda^T H_1(x(\tau))x^{\widetilde{\Delta}^2}(\tau) + x^{\widetilde{\Delta}}(\tau)^T (\lambda^T H_2(x(\tau)))x^{\widetilde{\Delta}}(\tau)\right) h_1(\sigma(t^*),\sigma(\tau))\widetilde{\Delta}\tau \\
&\quad + f^{\Delta\sigma_1\ldots\sigma_n}(x^*)(\widetilde{\sigma}(t^*)-t^*) \\
&\quad + \int_{t^*}^{\widetilde{\sigma}(t^*)} \left(F_1(x(\tau))x^{\widetilde{\Delta}^2}(\tau) + x^{\widetilde{\Delta}}(\tau)^T F_2(x(\tau))x^{\widetilde{\Delta}}(\tau)\right) h_1(\widetilde{\sigma}(t^*),\sigma(\tau))\widetilde{\Delta}\tau \\
&= \left(f^{\Delta\sigma_1\ldots\sigma_n}(x^*) + \lambda^T h^{\Delta\sigma_1\ldots\sigma_n}(x^*)\right)(\widetilde{\sigma}(t^*)-t^*) \\
&\quad + \int_{t^*}^{\widetilde{\sigma}(t^*)} \Big(\left(F_1(x(\tau)) + \lambda^T H_1(x(\tau))\right)x^{\widetilde{\Delta}^2}(\tau) \\
&\quad + x^{\widetilde{\Delta}}(\tau)^T \left(F_2(x(\tau)) + \lambda^T H_2(x(\tau))\right)x^{\widetilde{\Delta}}(\tau) \Big) h_1(\widetilde{\sigma}(t^*),\sigma(\tau))\widetilde{\Delta}\tau \\
&\geq 0.
\end{aligned}
$$

This is a contradiction.

2. *Let x^* be right-dense. Then there is a sequence $\{y^k\}_{k\in\mathbb{N}}$ such that $y^k \to x^*$, as $k\to\infty$, $y^k = x(t^k)$, $t^k > t^*$, $k\in\mathbb{N}$, and y^k, $k\in\mathbb{N}$, are feasible, and $y^k > x^*$, $k\in\mathbb{N}$, and*

$$f(y^k) < f(x^*), \quad k\in\mathbb{N}.$$

We have

$$
\begin{aligned}
h_j(y^k) &= 0, \\
h_j(x^*) &= 0, \quad j\in\{1,\ldots,m\}.
\end{aligned}
$$

Then

$$0 = h_j(y^k) - h_j(x^*), \quad j\in\{1,\ldots,m\}.$$

On the other hand, by the Taylor formula, we get

$$\begin{aligned} 0 &= h_j(y^k)-h_j(x^*) \\ &= h_j^{\Delta\sigma_1\ldots\sigma_n}(x^*)(t^k-t^*) \\ &\quad+\int\limits_{t^*}^{t^k}\Big(H_{1j}(x(\tau))x^{\widetilde{\Delta}^2}(\tau)+x^{\widetilde{\Delta}}(\tau)^T H_{2j}(x(\tau))x^{\widetilde{\Delta}}(\tau)\Big)h_1(t^k,\sigma(\tau))\widetilde{\Delta}\tau, \end{aligned}$$

$j\in\{1,\ldots,m\}$. *We multiply by* λ_j, $j\in\{1,\ldots,m\}$, *and we find*

$$\begin{aligned} 0 &= \lambda_j h_j^{\Delta\sigma_1\ldots\sigma_n}(x^*)(t^k-t^*) \\ &\quad+\int\limits_{t^*}^{t^k}\Big(\lambda_j H_{1j}(x(\tau))x^{\widetilde{\Delta}^2}(\tau)+\lambda_j x^{\widetilde{\Delta}}(\tau)^T H_{2j}(x(\tau))x^{\widetilde{\Delta}}(\tau)\Big)h_1(t^k,\sigma(\tau))\widetilde{\Delta}\tau, \end{aligned}$$

$j\in\{1,\ldots,m\}$. *Therefore*

$$\begin{aligned} 0 &= \lambda^T h^{\Delta\sigma_1\ldots\sigma_n}(x^*)(t^k-t^*) \\ &\quad+\int\limits_{t^*}^{t^k}\Big(\lambda^T H_1(x(\tau))x^{\widetilde{\Delta}^2}(\tau)+x^{\widetilde{\Delta}}(\tau)^T(\lambda^T H_2(x(\tau)))x^{\widetilde{\Delta}}(\tau)\Big)h_1(t^k,\sigma(\tau))\widetilde{\Delta}\tau. \end{aligned} \tag{5.17}$$

Now, we apply the Taylor formula for f *and we find*

$$\begin{aligned} 0 &> f(y^k)-f(x^*) \\ &= f^{\Delta\sigma_1\ldots\sigma_n}(x^*)(t^k-t^*) \\ &\quad+\int\limits_{t^*}^{t^k}\Big(F_1(x(\tau))x^{\widetilde{\Delta}^2}(\tau)+x^{\widetilde{\Delta}}(\tau)^T F_2(x(\tau))x^{\widetilde{\Delta}}(\tau)\Big)h_1(t^k,\sigma(\tau))\widetilde{\Delta}\tau. \end{aligned}$$

We add the last inequality and (5.17) *and we arrive at*

$$\begin{aligned} 0 &> \lambda^T h^{\Delta\sigma_1\ldots\sigma_n}(x^*)(t^k-t^*) \\ &\quad+\int\limits_{t^*}^{t^k}\Big(\lambda^T H_1(x(\tau))x^{\widetilde{\Delta}^2}(\tau)+x^{\widetilde{\Delta}}(\tau)^T(\lambda^T H_2(x(\tau)))x^{\widetilde{\Delta}}(\tau)\Big)h_1(t^k,\sigma(\tau))\widetilde{\Delta}\tau \end{aligned}$$

$$
\begin{aligned}
&+f^{\Delta\sigma_1\ldots\sigma_n}(x^*)(t^k-t^*)\\
&+\int_{t^*}^{t^k}\Big(F_1(x(\tau))x^{\widetilde{\Delta}^2}(\tau)+x^{\widetilde{\Delta}}(\tau)^T F_2(x(\tau))x^{\widetilde{\Delta}}(\tau)\Big)h_1(t^k,\sigma(\tau))\widetilde{\Delta}\tau\\
=\;&\Big(f^{\Delta\sigma_1\ldots\sigma_n}(x^*)+\lambda^T h^{\Delta\sigma_1\ldots\sigma_n}(x^*)\Big)(t^k-t^*)\\
&+\int_{t^*}^{t^k}\Bigg(\Big(F_1(x(\tau))+\lambda^T H_1(x(\tau))\Big)x^{\widetilde{\Delta}^2}(\tau)\\
&+x^{\widetilde{\Delta}}(\tau)^T\Big(F_2(x(\tau))+\lambda^T H_2(x(\tau))\Big)x^{\widetilde{\Delta}}(\tau)\Bigg)h_1(t^k,\sigma(\tau))\widetilde{\Delta}\tau\\
\geq\;&0.
\end{aligned}
$$

This is a contradiction.

3. *Let x^* be left-scattered and $\rho(x^*)$ is feasible, and*

$$f(\rho(x^*))<f(x^*).$$

Let also, $\rho(x^*)=x(t_2)$ for some $t_2<t^*$. We have

$$
\begin{aligned}
h_j(\rho(x^*)) &= 0,\\
h_j(x^*) &= 0,\quad j\in\{1,\ldots,m\}.
\end{aligned}
$$

Then

$$
\begin{aligned}
0 &= h_j(\rho(x^*))-h_j(x^*)\\
&= h_j^{\nabla\rho_1\ldots\rho_n}(x^*)(\rho(x^*)-x^*),\quad j\in\{1,\ldots,m\}.
\end{aligned}
$$

On the other hand, by the Taylor formula, we get

$$
\begin{aligned}
0 &= h_j(\rho(x^*))-h_j(x^*)\\
&= h_j^{\nabla\rho_1\ldots\rho_n}(x^*)(\widetilde{\rho}(t^*)-t^*)\\
&+\int_{t^*}^{\widetilde{\rho}(t^*)}\Big(H_{1j}^1(x(\tau))x^{\widetilde{\nabla}^2}(\tau)+x^{\widetilde{\nabla}}(\tau)^T H_{2j}^1(x(\tau))x^{\widetilde{\nabla}}(\tau)\Big)h_1(\widetilde{\rho}(t^*),\widetilde{\rho}(\tau))\widetilde{\nabla}\tau,
\end{aligned}
$$

$j \in \{1, \ldots, m\}$. *We multiply by* λ_j, $j \in \{1, \ldots, m\}$, *and we find*

$$\begin{aligned} 0 &= \lambda_j h_j^{\nabla \rho_1 \ldots \rho_n}(x^*)(\widetilde{\rho}(t^*) - t^*) \\ &+ \int_{t^*}^{\widetilde{\rho}(t^*)} \left(\lambda_j H_{1j}^1(x(\tau)) x^{\widetilde{\nabla}^2}(\tau) + \lambda_j x^{\widetilde{\nabla}}(\tau)^T H_{2j}^1(x(\tau)) x^{\widetilde{\nabla}}(\tau) \right) h_1(\widetilde{\rho}(t^*), \rho(\tau)) \widetilde{\nabla}\tau, \end{aligned}$$

$j \in \{1, \ldots, m\}$. *Therefore*

$$\begin{aligned} 0 &= \lambda^T h^{\nabla \rho_1 \ldots \rho_n}(x^*)(\widetilde{\rho}(t^*) - t^*) \\ &+ \int_{t^*}^{\widetilde{\rho}(t^*)} \left(\lambda^T H_1^1(x(\tau)) x^{\widetilde{\nabla}^2}(\tau) + x^{\widetilde{\nabla}}(\tau)^T (\lambda^T H_2^1(x(\tau))) x^{\widetilde{\nabla}}(\tau) \right) h_1(\widetilde{\rho}(t^*), \rho(\tau)) \widetilde{\nabla}\tau. \end{aligned} \tag{5.18}$$

Now, we apply the Taylor formula for f *and we find*

$$\begin{aligned} 0 &> f(\rho(x^*)) - f(x^*) \\ &= f^{\nabla \rho_1 \ldots \rho_n}(x^*)(\widetilde{\rho}(t^*) - t^*) \\ &+ \int_{t^*}^{\widetilde{\rho}(t^*)} \left(F_1^1(x(\tau)) x^{\widetilde{\nabla}^2}(\tau) + x^{\widetilde{\nabla}}(\tau)^T F_2^1(x(\tau)) x^{\widetilde{\nabla}}(\tau) \right) h_1(\widetilde{\rho}(t^*), \rho(\tau)) \widetilde{\nabla}\tau. \end{aligned}$$

We add the last inequality and (5.18) *and we arrive at*

$$\begin{aligned} 0 &> \lambda^T h^{\nabla \rho_1 \ldots \rho_n}(x^*)(\widetilde{\rho}(t^*) - t^*) \\ &+ \int_{t^*}^{\widetilde{\rho}(t^*)} \left(\lambda^T H_1^1(x(\tau)) x^{\widetilde{\nabla}^2}(\tau) + x^{\widetilde{\nabla}}(\tau)^T (\lambda^T H_2^1(x(\tau))) x^{\widetilde{\nabla}}(\tau) \right) h_1(\widetilde{\rho}(t^*), \rho(\tau)) \widetilde{\nabla}\tau \\ &+ f^{\nabla \rho_1 \ldots \rho_n}(x^*)(\widetilde{\rho}(t^*) - t^*) \\ &+ \int_{t^*}^{\widetilde{\rho}(t^*)} \left(F_1^1(x(\tau)) x^{\widetilde{\nabla}^2}(\tau) + x^{\widetilde{\nabla}}(\tau)^T F_2(x(\tau)) x^{\widetilde{\nabla}}(\tau) \right) h_1(\widetilde{\rho}(t^*), \rho(\tau)) \widetilde{\nabla}\tau \\ &= \left(f^{\nabla \rho_1 \ldots \rho_n}(x^*) + \lambda^T h^{\nabla \rho_1 \ldots \rho_n}(x^*) \right) (\widetilde{\rho}(t^*) - t^*) \\ &+ \int_{t^*}^{\widetilde{\rho}(t^*)} \Bigg(\left(F_1^1(x(\tau)) + \lambda^T H_1^1(x(\tau)) \right) x^{\widetilde{\nabla}^2}(\tau) \end{aligned}$$

$$+x^{\widetilde{\nabla}}(\tau)^T\left(F_2^1(x(\tau))+\lambda^T H_2^1(x(\tau))\right)x^{\widetilde{\nabla}}(\tau)\Big)h_1(\widetilde{\rho}(t^*),\rho(\tau))\widetilde{\nabla}\tau$$

$$\geq \quad 0.$$

This is a contradiction.

4. *Let x^* be left-dense. Then there is a sequence $\{y^k\}_{k\in\mathbb{N}}$ such that $y^k \to x^*$, as $k\to\infty$, $y^k = x(t^k)$, $t^k < t^*$, $k\in\mathbb{N}$, and y^k, $k\in\mathbb{N}$, are feasible, and $y^k < x^*$, $k\in\mathbb{N}$, and*

$$f(y^k) < f(x^*), \quad k\in\mathbb{N}.$$

We have

$$\begin{aligned} h_j(y^k) &= 0, \\ h_j(x^*) &= 0, \quad j\in\{1,\ldots,m\}. \end{aligned}$$

Then

$$0 = h_j(y^k)-h_j(x^*), \quad j\in\{1,\ldots,m\}.$$

On the other hand, by the Taylor formula, we get

$$\begin{aligned} 0 &= h_j(y^k)-h_j(x^*) \\ &= h_j^{\nabla\rho_1\ldots\rho_n}(x^*)(t^k-t^*) \\ &\quad +\int_{t^*}^{t^k}\left(H_{1j}^1(x(\tau))x^{\widetilde{\nabla}^2}(\tau)+x^{\widetilde{\nabla}}(\tau)^T H_{2j}^1(x(\tau))x^{\widetilde{\nabla}}(\tau)\right)h_1(t^k,\rho(\tau))\widetilde{\nabla}\tau, \end{aligned}$$

$j\in\{1,\ldots,m\}$. We multiply by λ_j, $j\in\{1,\ldots,m\}$, and we find

$$\begin{aligned} 0 &= \lambda_j h_j^{\nabla\rho_1\ldots\rho_n}(x^*)(t^k-t^*) \\ &\quad +\int_{t^*}^{t^k}\left(\lambda_j H_{1j}^1(x(\tau))x^{\widetilde{\nabla}^2}(\tau)+\lambda_j x^{\widetilde{\nabla}}(\tau)^T H_{2j}^1(x(\tau))x^{\widetilde{\nabla}}(\tau)\right)h_1(t^k,\rho(\tau))\widetilde{\nabla}\tau, \end{aligned}$$

$j\in\{1,\ldots,m\}$. Therefore

$$\begin{aligned} 0 &= \lambda^T h^{\nabla\rho_1\ldots\rho_n}(x^*)(t^k-t^*) \\ &\quad +\int_{t^*}^{t^k}\left(\lambda^T H_1^1(x(\tau))x^{\widetilde{\nabla}^2}(\tau)+x^{\widetilde{\nabla}}(\tau)^T(\lambda^T H_2^1(x(\tau)))x^{\widetilde{\nabla}}(\tau)\right)h_1(t^k,\rho(\tau))\widetilde{\nabla}\tau. \end{aligned} \tag{5.19}$$

Now, we apply the Taylor formula for f and we find

$$\begin{aligned} 0 &> f(y^k) - f(x^*) \\ &= f^{\nabla\rho_1\ldots\rho_n}(x^*)(t^k - t^*) \\ &\quad + \int_{t^*}^{t^k} \left(F_1^1(x(\tau)) x^{\tilde{\nabla}^2}(\tau) + x^{\tilde{\nabla}}(\tau)^T F_2^1(x(\tau)) x^{\tilde{\nabla}}(\tau) \right) h_1(t^k, \rho(\tau)) \tilde{\nabla}\tau. \end{aligned}$$

We add the last inequality and (5.19) *and we arrive at*

$$\begin{aligned} 0 &> \lambda^T h^{\nabla\rho_1\ldots\rho_n}(x^*)(t^k - t^*) \\ &\quad + \int_{t^*}^{t^k} \left(\lambda^T H_1^1(x(\tau)) x^{\tilde{\nabla}^2}(\tau) + x^{\tilde{\nabla}}(\tau)^T (\lambda^T H_2(x(\tau))) x^{\tilde{\nabla}}(\tau) \right) h_1(t^k, \rho(\tau)) \tilde{\nabla}\tau \\ &\quad + f^{\nabla\rho_1\ldots\rho_n}(x^*)(t^k - t^*) \\ &\quad + \int_{t^*}^{t^k} \left(F_1^1(x(\tau)) x^{\tilde{\nabla}^2}(\tau) + x^{\tilde{\nabla}}(\tau)^T F_2^1(x(\tau)) x^{\tilde{\nabla}}(\tau) \right) h_1(t^k, \rho(\tau)) \tilde{\nabla}\tau \\ &= \left(f^{\nabla\rho_1\ldots\rho_n}(x^*) + \lambda^T h^{\nabla\rho_1\ldots\rho_n}(x^*) \right)(t^k - t^*) \\ &\quad + \int_{t^*}^{t^k} \Bigg(\left(F_1^1(x(\tau)) + \lambda^T H_1^1(x(\tau)) \right) x^{\tilde{\nabla}^2}(\tau) \\ &\quad + x^{\tilde{\nabla}}(\tau)^T \left(F_2^1(x(\tau)) + \lambda^T H_2^1(x(\tau)) \right) x^{\tilde{\nabla}}(\tau) \Bigg) h_1(t^k, \rho(\tau)) \tilde{\nabla}\tau \\ &\geq 0. \end{aligned}$$

This is a contradiction. This completes the proof.

As above, one can prove the following results regarding the local maximum points of the objective function.

Theorem 40 *(Second Order Sufficiency Conditions) Suppose that there is a point x^* such that*

$$h(x^*) = 0$$

and a $\lambda \in \mathbb{R}^m$ *such that*

$$
\begin{aligned}
f^{\Delta\sigma_1\ldots\sigma_n}(x^*) + \lambda^T h^{\Delta\sigma_1\ldots\sigma_n}(x^*) &\leq 0,\\
f^{\nabla\rho_1\ldots\rho_n}(x^*) + \lambda^T h^{\nabla\rho_1\ldots\rho_n}(x^*) &\geq 0,\\
F_1(x^*) + \lambda^T H_1(x^*) &\leq 0,\\
F_1^1(x^*) + \lambda^T H_1^1(x^*) &\geq 0.
\end{aligned}
$$

Suppose also, that

$$L(x^*) = F_2(x^*) + \lambda^T H_2(x^*)$$

and

$$L^1(x^*) = F_2^1(x^*) + \lambda^T H_2^1(x^*)$$

are negative definite on

$$M = \{y \in \Lambda^m : h^{\Delta\sigma_1\ldots\sigma_n}(x^*)y = 0\}$$

and

$$M^1 = \{y \in \Lambda^m : h^{\nabla\rho_1\ldots\rho_n}(x^*)y = 0\},$$

respectively, that is

$$y^T L(x^*)y < 0$$

and

$$z^T L^1(x^*)z < 0,$$

respectively, for any $y \in M$, $y \neq 0$ *and* $z \in M^1$, $z \neq 0$. *Then* x^* *is a local maximum point for the objective function* f *subject to*

$$h(x) = 0.$$

5. Inequality Constraints

In this section, we consider a new problem

$$
\begin{aligned}
&\text{minimize} \quad f(x)\\
&\text{subject to} \quad h(x) = 0, \quad g(x) \leq 0,
\end{aligned}
\tag{5.20}
$$

where f and h are as before and g is a p dimensional function. Initially suppose that f, g and h are continuously Δ and ∇-differentiable.

Definition 23 *Let x^* be a point such that*

$$h(x^*) = 0, \quad g(x^*) \le 0, \tag{5.21}$$

and let J be the set of indices j for which

$$g_j(x^*) = 0.$$

The point x^* is said to be a Δ-regular point of the constraints (5.21) *if*

$$h_i^{\Delta\sigma_1\ldots\sigma_n}(x^*), \quad g_j^{\Delta\sigma_1\ldots\sigma_n}(x^*), \quad i \in \{1,\ldots,m\}, \quad j \in J,$$

are linearly independent.

The point x^* is said to be a ∇-regular point of the constraints (5.21) *if*

$$h_i^{\nabla\rho_1\ldots\rho_n}(x^*), \quad g_j^{\nabla\rho_1\ldots\rho_n}(x^*), \quad i \in \{1,\ldots,m\}, \quad j \in J,$$

are linearly independent.

Theorem 41 *(Second Order Necessary Conditions) Let x^* be a local minimum point for the problem* (5.20) *and suppose that x^* is a Δ-regular and ∇-regular point. Then there are vectors $\lambda \in \mathbb{R}^m$ and $\theta \in \mathbb{R}^p$ such that*

$$\begin{aligned}
f^{\Delta\sigma_1\ldots\sigma_n}(x^*) + \lambda^T h^{\Delta\sigma_1\ldots\sigma_n}(x^*) + \theta^T g^{\Delta\sigma_1\ldots\sigma_n}(x^*) &\ge 0, \\
f^{\nabla\rho_1\ldots\rho_n}(x^*) + \lambda^T h^{\nabla\rho_1\ldots\rho_n}(x^*) + \theta^T g^{\nabla\rho_1\ldots\rho_n}(x^*) &\le 0, \\
\theta^T g(x^*) &= 0.
\end{aligned} \tag{5.22}$$

Proof 30 *Note that, if $\theta \ge 0$ and $g(x^*) \le 0$, the condition*

$$\theta^T g(x^*) = 0$$

is equivalent to the statement that a component of θ may be nonzero only if the corresponding constraint is active. If $g_j(x^*) < 0$, then $\theta_j = 0$ and $\theta_j > 0$ implies $g_j(x^*) = 0$.

Because x^* is a local minimum point over the constraint set, it is also a local minimum point over the subset of that set defined by setting the active constraints to zero. Thus, for the resulting equality constrained problem defined in a neighbourhood of x^*, there are Lagrange multipliers. Therefore the first and second conditions of (5.22) *hold with $\theta_j = 0$ if $g_j(x^*) = 0$.*

Now, we will prove that $\theta \geq 0$. Assume that $\theta_k < 0$ for some $k \in J$. Let S and M be the surface and tangent plane, respectively, defined by all other active constraints at x^*. Since x^* is a Δ-regular point, there is an $y \in M$ such that

$$g^{\Delta\sigma_1\ldots\sigma_n}(x^*)y < 0.$$

Let $x(t)$ be a curve on S passing through x^* at t^* with

$$x^{\tilde{\Delta}}(t^*) = y.$$

Then, for small $t \geq t^*$, $x(t)$ is feasible and

$$\begin{aligned} f^{\tilde{\Delta}}(x^*) &= f^{\Delta\sigma_1\ldots\sigma_n}(x^*)y \\ &< 0. \end{aligned}$$

This is a contradiction. This completes the proof.

Example 5 *Let $\mathbb{T}_1 = \mathbb{T}_2 = \mathbb{Z}$. Consider the problem*

$$\textit{minimize} \quad 2x_1^2 + 2x_1x_2 + x_2^2 - 10x_1 - 10x_2$$

$$\textit{subject to} \quad x_1^2 + x_2^2 \leq 10, \quad 3x_1 + x_2 \leq 6.$$

Here

$$\begin{aligned} \sigma_1(x_1) &= x_1 + 1, \quad x_1 \in \mathbb{T}_1, \\ \sigma_2(x_2) &= x_2 + 1, \quad x_2 \in \mathbb{T}_2, \\ \rho_1(x_1) &= x_1 - 1, \quad x_1 \in \mathbb{T}_1, \\ \rho_2(x_2) &= x_2 - 1, \quad x_2 \in \mathbb{T}_2, \end{aligned}$$

and

$$\begin{aligned} f(x_1, x_2) &= 2x_1^2 + 2x_1x_2 + x_2^2 - 10x_1 - 10x_2, \\ h(x_1, x_2) &= 0, \end{aligned}$$

$$\begin{aligned} g_1(x_1,x_2) &= x_1^2+x_2^2-10, \\ g_2(x_1,x_2) &= 3x_1+x_2-6, \quad (x_1,x_2)\in\Lambda^2. \end{aligned}$$

Then

$$\begin{aligned} f_{x_1}^{\Delta_1}(x_1,x_2) &= 2(\sigma_1(x_1)+x_1)+2x_2-10 \\ &= 2(2x_1+1)+2x_2-10 \\ &= 4x_1+2x_2-8, \\ f_{x_2}^{\Delta_2\sigma_1}(x_1,x_2) &= 2\sigma_1(x_1)+x_2+x_2-10 \\ &= 2(x_1+1)+x_2+x_2-10 \\ &= 2x_1+2x_2-8, \quad (x_1,x_2)\in\Lambda^2, \end{aligned}$$

and

$$\begin{aligned} g_{1x_1}^{\Delta_1}(x_1,x_2) &= \sigma_1(x_1)+x_1 \\ &= x_1+1+x_1 \\ &= 2x_1+1, \\ g_{1x_2}^{\Delta_2\sigma_1}(x_1,x_2) &= \sigma_2(x_2)+x_2 \\ &= x_2+1+x_2 \\ &= 2x_2+1, \\ g_{2x_1}^{\Delta_1}(x_1,x_2) &= 3, \\ g_{2x_2}^{\Delta_2\sigma_1}(x_1,x_2) &= 1, \quad (x_1,x_2)\in\Lambda^2, \end{aligned}$$

and

$$f_{x_1}^{\nabla_1}(x_1,x_2) = 2(\rho_1(x_1)+x_1)+2x_2-10$$

$$
\begin{aligned}
&= 2(2x_1-1)+2x_2-10\\
&= 4x_1+2x_2-12,\\
f_{x_2}^{\nabla_2\rho_1}(x_1,x_2) &= 2\rho_1(x_1)+x_2+x_2-10\\
&= 2(x_1-1)+x_2+x_2-10\\
&= 2x_1+2x_2-12, \quad (x_1,x_2)\in\Lambda^2,
\end{aligned}
$$

and

$$
\begin{aligned}
g_{1x_1}^{\nabla_1}(x_1,x_2) &= \rho_1(x_1)+x_1\\
&= x_1-1+x_1\\
&= 2x_1-1,\\
g_{1x_2}^{\nabla_2\rho_1}(x_1,x_2) &= \rho_2(x_2)+x_2\\
&= x_2-1+x_2\\
&= 2x_2-1,\\
g_{2x_1}^{\nabla_1}(x_1,x_2) &= 3,\\
g_{2x_2}^{\nabla_2\rho_1}(x_1,x_2) &= 1, \quad (x_1,x_2)\in\Lambda^2.
\end{aligned}
$$

Then

$$
\begin{aligned}
&f^{\Delta\sigma_1\sigma_2}(x)+\lambda^T h^{\Delta\sigma_1\sigma_2}(x)+\theta^T g^{\Delta\sigma_1\sigma_2}(x)\\
&= (4x_1+2x_2-8, 2x_1+2x_2-8)+(\theta_1,\theta_2)\begin{pmatrix} 2x_1+1 & 2x_2+1\\ 3 & 1\end{pmatrix}\\
&= (4x_1+2x_2-8, 2x_1+2x_2-8)+(\theta_1(2x_1+1)+3\theta_2, \theta_1(2x_2+1)+\theta_2)
\end{aligned}
$$

and

$$
\begin{aligned}
& f^{\nabla \rho_1 \rho_2}(x) + \lambda^T h^{\nabla \rho_1 \rho_2}(x) + \theta^T g^{\nabla \rho_1 \rho_2}(x) \\
= \; & (4x_1 + 2x_2 - 12, 2x_1 + 2x_2 - 12) + (\theta_1, \theta_2) \begin{pmatrix} 2x_1 - 1 & 2x_2 - 1 \\ 3 & 1 \end{pmatrix} \\
= \; & (4x_1 + 2x_2 - 12, 2x_1 + 2x_2 - 12) + (\theta_1(2x_1 - 1) + 3\theta_2, \theta_1(2x_2 - 1) + \theta_2).
\end{aligned}
$$

Thus, we get the system

$$
\begin{aligned}
4x_1 + 2x_2 - 8 + \theta_1(2x_1 + 1) + 3\theta_2 &\geq 0 \\
2x_1 + 2x_2 - 8 + \theta_1(2x_1 + 1) + \theta_2 &\geq 0 \\
4x_1 + 2x_2 - 12 + \theta_1(2x_1 - 1) + 3\theta_2 &\leq 0 \\
2x_1 + 2x_2 - 12 + \theta_1(2x_1 - 1) + \theta_2 &\leq 0 \\
\theta_1(x_1^2 + x_2^2 - 10) &= 0 \\
\theta_2(3x_1 + x_2 - 6) &= 0.
\end{aligned}
$$

To find a solution, we define various of combinations of active constraints. Assuming the first constraint is not active and the second constraint is active, we obtain

$$
\begin{aligned}
4x_1 + 2x_2 - 8 + 3\theta_2 &\geq 0 \\
2x_1 + 2x_2 - 8 + \theta_2 &\geq 0 \\
4x_1 + 2x_2 - 12 + 3\theta_2 &\leq 0 \\
2x_1 + 2x_2 - 12 + \theta_2 &\leq 0 \\
(3x_1 + x_2 - 6) &= 0.
\end{aligned}
$$

We have that

$$
x_1 = 1,
$$

$$
\begin{aligned}
x_2 &= 3, \\
\theta_1 &= 0, \\
\theta_2 &= \frac{1}{2}
\end{aligned}
$$

is its solution. Observe that

$$x_1^2 + x_2^2 = 10,$$

i.e., the first constraint holds.

Let

$$G_1(x(t)) = \left(g^{\Delta\sigma_1\ldots\sigma_n}(x(t))\right)^{\sigma_1\ldots\sigma_n}$$

and

$$G_{2j}(x(t)) = \begin{pmatrix} g_{jx_1x_1}^{\Delta_1^2}(x(t)) & g_{jx_1x_2}^{\Delta_1\Delta_2\sigma_1}(x(t)) & \cdots & g_{jx_1x_n}^{\Delta_1\Delta_n\sigma_1\ldots\sigma_n}(x(t)) \\ g_{jx_2x_1}^{\Delta_2\sigma_1\Delta_1}(x(t)) & g_{jx_2x_2}^{\Delta_2\sigma_1\Delta_2\sigma_1}(x(t)) & \cdots & g_{jx_2x_n}^{\Delta_2\sigma_1\Delta_n\sigma_1\ldots\sigma_{n-1}}(x(t)) \\ \vdots & \vdots & \vdots & \vdots \\ g_{jx_nx_1}^{\Delta_n\sigma_1\ldots\sigma_{n-1}\Delta_1}(x(t)) & g_{jx_nx_2}^{\Delta_n\sigma_1\ldots\sigma_{n-1}\Delta_2\sigma_1}(x(t)) & \cdots & g_{jx_nx_n}^{\Delta_n\sigma_1\ldots\sigma_{n-1}\Delta_n\sigma_1\ldots\sigma_{n-1}}(x(t)) \end{pmatrix},$$

$j \in \{1, \ldots, p\}$, and

$$G_2(x(t)) = \begin{pmatrix} G_{21}(x(t)) \\ G_{22}(x(t)) \\ \vdots \\ G_{2p}(x(t)) \end{pmatrix}.$$

Then

$$g^{\tilde{\Delta}^2}(t) = G_1(x(t))x^{\tilde{\Delta}}(t) + x^{\tilde{\Delta}}(t)^T G_2(x(t))x^{\tilde{\Delta}}(t).$$

Let also,

$$G_1^1(x(t)) = \left(g^{\nabla\rho_1\ldots\rho_n}(x(t))\right)^{\rho_1\ldots\rho_n}$$

and

$$G_{2j}^1(x(t)) = \begin{pmatrix} g_{jx_1x_1}^{\nabla_1^2}(x(t)) & g_{jx_1x_2}^{\nabla_1\nabla_2\rho_1}(x(t)) & \cdots & g_{jx_1x_n}^{\nabla_1\nabla_n\rho_1\ldots\rho_n}(x(t)) \\ g_{jx_2x_1}^{\nabla_2\rho_1\nabla_1}(x(t)) & g_{jx_2x_2}^{\nabla_2\rho_1\nabla_2\rho_1}(x(t)) & \cdots & g_{jx_2x_n}^{\nabla_2\rho_1\nabla_n\rho_1\ldots\rho_{n-1}}(x(t)) \\ \vdots & \vdots & \vdots & \vdots \\ g_{jx_nx_1}^{\nabla_n\rho_1\ldots\rho_{n-1}\nabla_1}(x(t)) & g_{jx_nx_2}^{\nabla_n\rho_1\ldots\rho_{n-1}\nabla_2\rho_1}(x(t)) & \cdots & g_{jx_nx_n}^{\nabla_n\rho_1\ldots\rho_{n-1}\nabla_n\rho_1\ldots\rho_{n-1}}(x(t)) \end{pmatrix},$$

$j \in \{1, \ldots, p\}$, and

$$G_2^1(x(t) = \begin{pmatrix} G_{21}^1(x(t)) \\ G_{22}^1(x(t)) \\ \vdots \\ G_{2m}^1(x(t)) \end{pmatrix}.$$

Then, we have

$$g^{\widetilde{\nabla}^2}(t) = G_1^1(x(t))x^{\widetilde{\nabla}}(t) + x^{\widetilde{\nabla}}(t)^T G_2^1(x(t))x^{\widetilde{\nabla}}(t).$$

Theorem 42 *(Second Order Necessary Conditions) Suppose that x^* is a local minimum point of the objective function f subject to*

$$h(x) = 0, \quad g(x) \leq 0,$$

and x^* is a Δ-regular and ∇-regular point for these constraints. Then, there is a $\lambda \in \mathbb{R}^m$ and $\theta \in \mathbb{R}^p$ such that

$$F_1(x^*) + \lambda^T H_1(x^*)x^{\widetilde{\Delta}^2}(t^*) + \theta^T G_1(x^*)x^{\widetilde{\Delta}^2}(t^*) = 0. \tag{5.23}$$

If we denote by M the Δ-tangent plane of the active constraints of x^*, then the matrix

$$L(x^*) = F_2(x^*) + \lambda^T H_2(x^*) + \theta^T G_2(x^*)$$

is positive semidefinite on M, i.e.,

$$y^T L(x^*)y \geq 0 \quad \textit{for any} \quad y \in M.$$

Proof 31 *Note that, in particular x^* is a solution to the problem*

$$\textit{minimize} \quad f(x)$$

$$\textit{subject to} \quad h(x) = 0, \quad g(x) = 0.$$

Then the value of the linear problem

$$\textit{minimize} \quad F_1(x^*)y$$

$$\textit{subject to} \quad H_1(x^*)y = 0, \quad G_1(x^*) = 0$$

is zero. Then, applying the duality theorem, Theorem 30, the dual problem is feasible. Therefore there is a $\lambda \in \mathbb{R}^m$ and $\theta \in \mathbb{R}^p$ such that (45) *holds. Note that*

$$f^{\widetilde{\Delta}^2}(x^*) \geq 0.$$

By definition, we find

$$f^{\widetilde{\Delta}^2}(x^*) = F_1(x^*)x^{\widetilde{\Delta}^2}(t^*) + x^{\widetilde{\Delta}}(t^*)^T F_2(x^*)x^{\widetilde{\Delta}}(t^*). \tag{5.24}$$

If we $\widetilde{\Delta}$-differentiate the relation

$$\lambda^T h(x(t)) = 0$$

twice, we obtain

$$\lambda^T H_1(x^*)x^{\widetilde{\Delta}^2}(t^*) + x^{\widetilde{\Delta}}(t^*)^T \lambda^T H_2(x^*)x^{\widetilde{\Delta}}(t^*) = 0. \tag{5.25}$$

If we $\widetilde{\Delta}$-differentiate the relation

$$\theta^T g(x(t)) = 0$$

twice, we obtain

$$\lambda^T G_1(x^*)x^{\widetilde{\Delta}^2}(t^*) + x^{\widetilde{\Delta}}(t^*)^T \lambda^T G_2(x^*)x^{\widetilde{\Delta}}(t^*) = 0. \tag{5.26}$$

Adding (5.24), (5.25) *and* (5.26), *we arrive at*

$$\begin{aligned}
0 &\leq F_1(x^*)x^{\widetilde{\Delta}^2}(t^*) + x^{\widetilde{\Delta}}(t^*)^T F_2(x^*)x^{\widetilde{\Delta}}(t^*) \\
&\quad + \lambda^T H_1(x^*)x^{\widetilde{\Delta}^2}(t^*) + x^{\widetilde{\Delta}}(t^*)^T \lambda^T H_2(x^*)x^{\widetilde{\Delta}}(t^*) \\
&\quad + \theta^T G_1(x^*)x^{\widetilde{\Delta}^2}(t^*) + x^{\widetilde{\Delta}}(t^*)^T \theta^T G_2(x^*)x^{\widetilde{\Delta}}(t^*) \\
&= \left(F_1(x^*) + \lambda^T H_1(x^*) + \theta^T G_1(x^*)\right) x^{\widetilde{\Delta}^2}(t^*) \\
&\quad + x^{\widetilde{\Delta}}(t^*)^T \left(F_2(x^*) + \lambda^T H_2(x^*) + \theta^T G_2(x^*)\right) x^{\widetilde{\Delta}}(t^*) \\
&= x^{\widetilde{\Delta}}(t^*)^T L(x^*)x^{\widetilde{\Delta}}(t^*) \\
&\geq 0.
\end{aligned}$$

Because $x^{\widetilde{\Delta}}(t^*)$ was arbitrarily chosen in M, the proof is completed.

As above, one can prove the following results.

Theorem 43 *(Second Order Necessary Conditions) Suppose that x^* is a local minimum point of the objective function f subject to*

$$h(x) = 0, \quad g(x) \leq 0,$$

and x^* is a Δ-regular and ∇-regular point for these constraints. Then, there is a $\lambda \in \mathbb{R}^m$ and $\theta \in \mathbb{R}^p$ such that

$$F_1^1(x^*) + \lambda^T H_1^1(x^*) x^{\widetilde{\Delta}^2}(t^*) + \theta^T G_1^1(x^*) x^{\widetilde{\Delta}^2}(t^*) = 0.$$

If we denote by M the ∇-tangent plane of the active constraints of x^*, then the matrix

$$L^1(x^*) = F_2^1(x^*) + \lambda^T H_2^1(x^*) + \theta^T G_2^1(x^*)$$

is positive semidefinite on M, i.e.,

$$y^T L^1(x^*) y \geq 0 \quad \textit{for any} \quad y \in M.$$

Theorem 44 *(Second Order Necessary Conditions) Suppose that x^* is a local maximum point of the objective function f subject to*

$$h(x) = 0, \quad g(x) \leq 0,$$

and x^* is a Δ-regular and ∇-regular point for these constraints. Then, there is a $\lambda \in \mathbb{R}^m$ and $\theta \in \mathbb{R}^p$ such that

$$F_1(x^*) + \lambda^T H_1(x^*) x^{\widetilde{\Delta}^2}(t^*) + \theta^T G_1(x^*) x^{\widetilde{\Delta}^2}(t^*) = 0.$$

If we denote by M the Δ-tangent plane of the active constraints of x^*, then the matrix

$$L(x^*) = F_2(x^*) + \lambda^T H_2(x^*) + \theta^T G_2(x^*)$$

is negative semidefinite on M, i.e.,

$$y^T L(x^*) y \leq 0 \quad \textit{for any} \quad y \in M.$$

Theorem 45 *(Second Order Necessary Conditions) Suppose that x^* is a local maximum point of the objective function f subject to*

$$h(x) = 0, \quad g(x) \leq 0,$$

and x^* is a Δ-regular and ∇-regular point for these constraints. Then, there is a $\lambda \in \mathbb{R}^m$ and $\theta \in \mathbb{R}^p$ such that

$$F_1^1(x^*) + \lambda^T H_1^1(x^*) x^{\widetilde{\Delta}^2}(t^*) + \theta^T G_1^1(x^*) x^{\widetilde{\Delta}^2}(t^*) = 0.$$

If we denote by M the ∇-tangent plane of the active constraints of x^*, then the matrix

$$L^1(x^*) = F_2^1(x^*) + \lambda^T H_2^1(x^*) + \theta^T G_2^1(x^*)$$

is negative semidefinite on M, i.e.,

$$y^T L^1(x^*) y \leq 0 \quad \text{for any} \quad y \in M.$$

Theorem 46 *(Second Order Sufficiency Conditions) Suppose that there is a point x^* such that*

$$h(x^*) = 0$$

and a $\lambda \in \mathbb{R}^m$ and $\theta \in \mathbb{R}^p$ such that

$$\begin{aligned}
\theta^T g(x^*) &= 0, \\
f^{\Delta\sigma_1\ldots\sigma_n}(x^*) + \lambda^T h^{\Delta\sigma_1\ldots\sigma_n}(x^*) + \theta^T g^{\Delta\sigma_1\ldots\sigma_n}(x^*) &\geq 0, \\
f^{\nabla\rho_1\ldots\rho_n}(x^*) + \lambda^T h^{\nabla\rho_1\ldots\rho_n}(x^*) + \theta^T g^{\nabla\rho_1\ldots\rho_n} &\leq 0, \\
F_1(x^*) + \lambda^T H_1(x^*) + \theta^T G_1(x^*) &\geq 0, \\
F_1^1(x^*) + \lambda^T H_1^1(x^*) + \theta^T G^1(x^*) &\leq 0.
\end{aligned}$$

Suppose also, that

$$L(x^*) = F_2(x^*) + \lambda^T H_2(x^*) + \theta^T G_2(x^*)$$

and

$$L^1(x^*) = F_2^1(x^*) + \lambda^T H_2^1(x^*) + \theta^T G_2^1(x^*)$$

are positive definite on

$$M = \{y \in \Lambda^n : h^{\Delta\sigma_1\ldots\sigma_n}(x^*) y = 0, \quad g^{\Delta\sigma_1\ldots\sigma_n}(x^*) y = 0\}$$

and

$$M^1 = \{y \in \Lambda^n : h^{\nabla\rho_1\ldots\rho_n}(x^*) y = 0, \quad g^{\nabla\rho_1\ldots\rho_n}(x^*) y = 0\},$$

respectively, that is

$$y^T L(x^*) y > 0$$

and

$$z^T L^1(x^*) z > 0,$$

respectively, for any $y \in M$, $y \neq 0$ and $z \in M^1$, $z \neq 0$. Then x^* is a local minimum point for the objective function f subject to

$$h(x) = 0, \quad g(x) \leq 0.$$

Proof 32 *Assume the contrary, i.e., assume that x^* is not a local minimum point of f.*

1. *Let x^* be right-scattered and $\sigma(x^*)$ is feasible, and*

$$f(\sigma(x^*)) < f(x^*).$$

Let also, $\sigma(x^*) = x(t_1)$ for some $t_1 > t^*$. We have

$$\begin{aligned} h_j(\sigma(x^*)) &= 0, \\ h_j(x^*) &= 0, \quad j \in \{1, \ldots, m\}. \end{aligned}$$

Then

$$\begin{aligned} 0 &= h_j(\sigma(x^*)) - h_j(x^*) \\ &= h_j^{\Delta\sigma_1\ldots\sigma_n}(x^*)(\sigma(x^*) - x^*), \quad j \in \{1, \ldots, m\}. \end{aligned}$$

On the other hand, by the Taylor formula, we get

$$\begin{aligned} 0 &= h_j(\sigma(x^*)) - h_j(x^*) \\ &= h_j^{\Delta\sigma_1\ldots\sigma_n}(x^*)(\widetilde{\sigma}(t^*) - t^*) \\ &+ \int_{t^*}^{\widetilde{\sigma}(t^*)} \left(H_{1j}(x(\tau))x^{\widetilde{\Delta}^2}(\tau) + x^{\widetilde{\Delta}}(\tau)^T H_{2j}(x(\tau))x^{\widetilde{\Delta}}(\tau)\right) h_1(\widetilde{\sigma}(t^*), \sigma(\tau))\widetilde{\Delta}\tau, \end{aligned}$$

$j \in \{1, \ldots, m\}$. We multiply by λ_j, $j \in \{1, \ldots, m\}$, and we find

$$\begin{aligned} 0 &= \lambda_j h_j^{\Delta\sigma_1\ldots\sigma_n}(x^*)(\widetilde{\sigma}(t^*) - t^*) \\ &+ \int_{t^*}^{\widetilde{\sigma}(t^*)} \left(\lambda_j H_{1j}(x(\tau))x^{\widetilde{\Delta}^2}(\tau) + \lambda_j x^{\widetilde{\Delta}}(\tau)^T H_{2j}(x(\tau))x^{\widetilde{\Delta}}(\tau)\right) h_1(\widetilde{\sigma}(t^*), \sigma(\tau))\widetilde{\Delta}\tau, \end{aligned}$$

$j \in \{1, \ldots, m\}$. Therefore

$$\begin{aligned} 0 &= \lambda^T h^{\Delta\sigma_1\ldots\sigma_n}(x^*)(\widetilde{\sigma}(t^*) - t^*) \\ &+ \int_{t^*}^{\widetilde{\sigma}(t^*)} \left(\lambda^T H_1(x(\tau))x^{\widetilde{\Delta}^2}(\tau) + x^{\widetilde{\Delta}}(\tau)^T (\lambda^T H_2(x(\tau)))x^{\widetilde{\Delta}}(\tau)\right) h_1(\widetilde{\sigma}(t^*), \sigma(\tau))\widetilde{\Delta}\tau. \end{aligned} \tag{5.27}$$

Now, we apply the Taylor formula for g and we find

$$\begin{aligned} 0 &\geq g_j(\sigma(x^*)) - g_j(x^*) \\ &= g_j^{\Delta\sigma_1\dots\sigma_n}(x^*)(\widetilde{\sigma}(t^*) - t^*) \\ &\quad + \int_{t^*}^{\widetilde{\sigma}(t^*)} \Big(G_{1j}(x(\tau))x^{\widetilde{\Delta}^2}(\tau) + x^{\widetilde{\Delta}}(\tau)^T G_{2j}(x(\tau))x^{\widetilde{\Delta}}(\tau) \Big) h_1(\widetilde{\sigma}(t^*), \sigma(\tau))\widetilde{\Delta}\tau, \end{aligned}$$

$j \in \{1, \dots, p\}$ *and then we multiply by* θ_j, $j \in \{1, \dots, p\}$, *and we find*

$$\begin{aligned} 0 &\geq \theta_j g_j^{\Delta\sigma_1\dots\sigma_n}(x^*)(\widetilde{\sigma}(t^*) - t^*) \\ &\quad + \int_{t^*}^{\widetilde{\sigma}(t^*)} \Big(\theta_j G_{1j}(x(\tau))x^{\widetilde{\Delta}^2}(\tau) + \theta_j x^{\widetilde{\Delta}}(\tau)^T G_{2j}(x(\tau))x^{\widetilde{\Delta}}(\tau) \Big) h_1(\widetilde{\sigma}(t^*), \sigma(\tau))\widetilde{\Delta}\tau, \end{aligned}$$

$j \in \{1, \dots, p\}$. *Consequently*

$$\begin{aligned} 0 &\geq \theta^T g^{\Delta\sigma_1\dots\sigma_n}(x^*)(\widetilde{\sigma}(t^*) - t^*) \\ &\quad + \int_{t^*}^{\widetilde{\sigma}(t^*)} \Big(\theta^T G_1(x(\tau))x^{\widetilde{\Delta}^2}(\tau) + x^{\widetilde{\Delta}}(\tau)^T (\theta^T G_2(x(\tau)))x^{\widetilde{\Delta}}(\tau) \Big) h_1(\widetilde{\sigma}(t^*), \sigma(\tau))\widetilde{\Delta}\tau. \end{aligned} \tag{5.28}$$

Now, we apply the Taylor formula for f *and we find*

$$\begin{aligned} 0 &> f(\sigma(x^*)) - f(x^*) \\ &= f^{\Delta\sigma_1\dots\sigma_n}(x^*)(\widetilde{\sigma}(t^*) - t^*) \\ &\quad + \int_{t^*}^{\widetilde{\sigma}(t^*)} \Big(F_1(x(\tau))x^{\widetilde{\Delta}^2}(\tau) + x^{\widetilde{\Delta}}(\tau)^T F_2(x(\tau))x^{\widetilde{\Delta}}(\tau) \Big) h_1(\widetilde{\sigma}(t^*), \sigma(\tau))\widetilde{\Delta}\tau. \end{aligned}$$

We add the last inequality, (5.27) *and* (5.28), *and we arrive at*

$$\begin{aligned} 0 &> \lambda^T h^{\Delta\sigma_1\dots\sigma_n}(x^*)(\widetilde{\sigma}(t^*) - t^*) \\ &\quad + \int_{t^*}^{\widetilde{\sigma}(t^*)} \Big(\lambda^T H_1(x(\tau))x^{\widetilde{\Delta}^2}(\tau) + x^{\widetilde{\Delta}}(\tau)^T (\lambda^T H_2(x(\tau)))x^{\widetilde{\Delta}}(\tau) \Big) h_1(\widetilde{\sigma}(t^*), \sigma(\tau))\widetilde{\Delta}\tau \\ &\quad + \theta^T g^{\Delta\sigma_1\dots\sigma_n}(x^*)(\widetilde{\sigma}(t^*) - t^*) \\ &\quad + \int_{t^*}^{\widetilde{\sigma}(t^*)} \Big(\theta^T G_1(x(\tau))x^{\widetilde{\Delta}^2}(\tau) + x^{\widetilde{\Delta}}(\tau)^T (\theta^T G_2(x(\tau)))x^{\widetilde{\Delta}}(\tau) \Big) h_1(\widetilde{\sigma}(t^*), \sigma(\tau))\widetilde{\Delta}\tau \end{aligned}$$

$$
\begin{aligned}
&+f^{\Delta\sigma_1\ldots\sigma_n}(x^*)(\widetilde{\sigma}(t^*)-t^*)\\
&+\int_{t^*}^{\widetilde{\sigma}(t^*)}\Big(F_1(x(\tau))x^{\widetilde{\Delta}^2}(\tau)+x^{\widetilde{\Delta}}(\tau)^T F_2(x(\tau))x^{\widetilde{\Delta}}(\tau)\Big)h_1(\widetilde{\sigma}(t^*),\sigma(\tau))\widetilde{\Delta}\tau\\
=\;&\Big(f^{\Delta\sigma_1\ldots\sigma_n}(x^*)+\lambda^T h^{\Delta\sigma_1\ldots\sigma_n}(x^*)+\theta^T h^{\Delta\sigma_1\ldots\sigma_n}(x^*)\Big)(\widetilde{\sigma}(t^*)-t^*)\\
&+\int_{t^*}^{\widetilde{\sigma}(t^*)}\Bigg(\big(F_1(x(\tau))+\lambda^T H_1(x(\tau))+\theta^T G_1(x(\tau))\big)x^{\widetilde{\Delta}^2}(\tau)\\
&+x^{\widetilde{\Delta}}(\tau)^T\big(F_2(x(\tau))+\lambda^T H_2(x(\tau))+\theta^T G_2(x(\tau))\big)x^{\widetilde{\Delta}}(\tau)\Bigg)h_1(\widetilde{\sigma}(t^*),\sigma(\tau))\widetilde{\Delta}\tau\\
\geq\;&0.
\end{aligned}
$$

This is a contradiction.

2. *Let x^* be right-dense. Then there is a sequence $\{y^k\}_{k\in\mathbb{N}}$ such that $y^k\to x^*$, as $k\to\infty$, $y^k=x(t^k)$, $t^k>t^*$, $k\in\mathbb{N}$, and y^k, $k\in\mathbb{N}$, are feasible, and $y^k>x^*$, $k\in\mathbb{N}$, and*

$$f(y^k)<f(x^*),\quad k\in\mathbb{N}.$$

We have

$$
\begin{aligned}
h_j(y^k) &= 0,\\
h_j(x^*) &= 0,\quad j\in\{1,\ldots,m\}.
\end{aligned}
$$

Then

$$0 = h_j(y^k)-h_j(x^*),\quad j\in\{1,\ldots,m\}.$$

On the other hand, by the Taylor formula, we get

$$
\begin{aligned}
0 &= h_j(y^k)-h_j(x^*)\\
&= h_j^{\Delta\sigma_1\ldots\sigma_n}(x^*)(t^k-t^*)\\
&\quad+\int_{t^*}^{t^k}\Big(H_{1j}(x(\tau))x^{\widetilde{\Delta}^2}(\tau)+x^{\widetilde{\Delta}}(\tau)^T H_{2j}(x(\tau))x^{\widetilde{\Delta}}(\tau)\Big)h_1(t^k,\sigma(\tau))\widetilde{\Delta}\tau,
\end{aligned}
$$

$j\in\{1,\ldots,m\}$. We multiply by λ_j, $j\in\{1,\ldots,m\}$, and we find

$$0 = \lambda_j h_j^{\Delta\sigma_1\ldots\sigma_n}(x^*)(t^k-t^*)$$

$$+\int_{t^*}^{t^k}\Big(\lambda_j H_{1j}(x(\tau))x^{\widetilde{\Delta}^2}(\tau)+\lambda_j x^{\widetilde{\Delta}}(\tau)^T H_{2j}(x(\tau))x^{\widetilde{\Delta}}(\tau)\Big)h_1(t^k,\sigma(\tau))\widetilde{\Delta}\tau,$$

$j\in\{1,\ldots,m\}$. *Therefore*

$$\begin{aligned} 0 &= \lambda^T h^{\Delta\sigma_1\ldots\sigma_n}(x^*)(t^k-t^*) \\ &\quad +\int_{t^*}^{t^k}\Big(\lambda^T H_1(x(\tau))x^{\widetilde{\Delta}^2}(\tau)+x^{\widetilde{\Delta}}(\tau)^T(\lambda^T H_2(x(\tau)))x^{\widetilde{\Delta}}(\tau)\Big)h_1(t^k,\sigma(\tau))\widetilde{\Delta}\tau. \end{aligned} \tag{5.29}$$

We apply the Taylor formula for g and we arrive at

$$\begin{aligned} 0 &\geq g_j(y^k)-g_j(x^*) \\ &= g_j^{\Delta\sigma_1\ldots\sigma_n}(x^*)(t^k-t^*) \\ &\quad +\int_{t^*}^{t^k}\Big(G_{1j}(x(\tau))x^{\widetilde{\Delta}^2}(\tau)+x^{\widetilde{\Delta}}(\tau)^T G_{2j}(x(\tau))x^{\widetilde{\Delta}}(\tau)\Big)h_1(t^k,\sigma(\tau))\widetilde{\Delta}\tau, \end{aligned}$$

$j\in\{1,\ldots,p\}$. *We multiply by θ_j, $j\in\{1,\ldots,p\}$, and we find*

$$\begin{aligned} 0 &\geq \theta_j g_j^{\Delta\sigma_1\ldots\sigma_n}(x^*)(t^k-t^*) \\ &\quad +\int_{t^*}^{t^k}\Big(\theta_j G_{1j}(x(\tau))x^{\widetilde{\Delta}^2}(\tau)+\theta_j x^{\widetilde{\Delta}}(\tau)^T G_{2j}(x(\tau))x^{\widetilde{\Delta}}(\tau)\Big)h_1(t^k,\sigma(\tau))\widetilde{\Delta}\tau, \end{aligned}$$

$j\in\{1,\ldots,p\}$. *Consequently*

$$\begin{aligned} 0 &\geq \theta^T g^{\Delta\sigma_1\ldots\sigma_n}(x^*)(t^k-t^*) \\ &\quad +\int_{t^*}^{t^k}\Big(\theta^T G_1(x(\tau))x^{\widetilde{\Delta}^2}(\tau)+x^{\widetilde{\Delta}}(\tau)^T(\theta^T G_2(x(\tau)))x^{\widetilde{\Delta}}(\tau)\Big)h_1(t^k,\sigma(\tau))\widetilde{\Delta}\tau. \end{aligned} \tag{5.30}$$

Now, we apply the Taylor formula for f and we find

$$\begin{aligned} 0 &> f(y^k)-f(x^*) \\ &= f^{\Delta\sigma_1\ldots\sigma_n}(x^*)(t^k-t^*) \end{aligned}$$

$$+\int_{t^*}^{t^k}\Big(F_1(x(\tau))x^{\widetilde{\Delta}^2}(\tau)+x^{\widetilde{\Delta}}(\tau)^T F_2(x(\tau))x^{\widetilde{\Delta}}(\tau)\Big)h_1(t^k,\sigma(\tau))\widetilde{\Delta}\tau.$$

We add the last inequality, (5.29) *and* (5.30)*, and we arrive at*

$$\begin{aligned}
0 &> \lambda^T h^{\Delta\sigma_1\ldots\sigma_n}(x^*)(t^k-t^*)\\
&+\int_{t^*}^{t^k}\Big(\lambda^T H_1(x(\tau))x^{\widetilde{\Delta}^2}(\tau)+x^{\widetilde{\Delta}}(\tau)^T(\lambda^T H_2(x(\tau)))x^{\widetilde{\Delta}}(\tau)\Big)h_1(t^k,\sigma(\tau))\widetilde{\Delta}\tau\\
&+\theta^T g^{\Delta\sigma_1\ldots\sigma_n}(x^*)(t^k-t^*)\\
&+\int_{t^*}^{t^k}\Big(\theta^T G_1(x(\tau))x^{\widetilde{\Delta}^2}(\tau)+x^{\widetilde{\Delta}}(\tau)^T(\theta^T G_2(x(\tau)))x^{\widetilde{\Delta}}(\tau)\Big)h_1(t^k,\sigma(\tau))\widetilde{\Delta}\tau\\
&+f^{\Delta\sigma_1\ldots\sigma_n}(x^*)(t^k-t^*)\\
&+\int_{t^*}^{t^k}\Big(F_1(x(\tau))x^{\widetilde{\Delta}^2}(\tau)+x^{\widetilde{\Delta}}(\tau)^T F_2(x(\tau))x^{\widetilde{\Delta}}(\tau)\Big)h_1(t^k,\sigma(\tau))\widetilde{\Delta}\tau\\
&= \Big(f^{\Delta\sigma_1\ldots\sigma_n}(x^*)+\lambda^T h^{\Delta\sigma_1\ldots\sigma_n}(x^*)+\theta^T g^{\Delta\sigma_1\ldots\sigma_n}(x^*)\Big)(t^k-t^*)\\
&+\int_{t^*}^{t^k}\Bigg(\Big(F_1(x(\tau))+\lambda^T H_1(x(\tau))+\theta^T G_1(x(\tau))\Big)x^{\widetilde{\Delta}^2}(\tau)\\
&+x^{\widetilde{\Delta}}(\tau)^T\Big(F_2(x(\tau))+\lambda^T H_2(x(\tau))+\theta^T G_2(x(\tau))\Big)x^{\widetilde{\Delta}}(\tau)\Bigg)h_1(t^k,\sigma(\tau))\widetilde{\Delta}\tau\\
&\geq 0.
\end{aligned}$$

This is a contradiction.

3. *Let x^* be left-scattered and $\rho(x^*)$ is feasible, and*

$$f(\rho(x^*)) < f(x^*).$$

Let also, $\rho(x^*) = x(t_2)$ for some $t_2 < t^*$. We have

$$h_j(\rho(x^*)) = 0,$$

$$h_j(x^*) \quad = \quad 0, \quad j \in \{1,\ldots,m\}.$$

Then

$$\begin{aligned} 0 &= h_j(\rho(x^*)) - h_j(x^*) \\ &= h_j^{\nabla\rho_1\ldots\rho_n}(x^*)(\rho(x^*) - x^*), \quad j \in \{1,\ldots,m\}. \end{aligned}$$

On the other hand, by the Taylor formula, we get

$$\begin{aligned} 0 &= h_j(\rho(x^*)) - h_j(x^*) \\ &= h_j^{\nabla\rho_1\ldots\rho_n}(x^*)(\widetilde{\rho}(t^*) - t^*) \\ &\quad + \int_{t^*}^{\widetilde{\rho}(t^*)} \left(H_{1j}^1(x(\tau))x^{\widetilde{\nabla}^2}(\tau) + x^{\widetilde{\nabla}}(\tau)^T H_{2j}^1(x(\tau))x^{\widetilde{\nabla}}(\tau)\right) h_1(\widetilde{\rho}(t^*),\widetilde{\rho}(\tau))\widetilde{\nabla}\tau, \end{aligned}$$

$j \in \{1,\ldots,m\}$. *We multiply by* λ_j, $j \in \{1,\ldots,m\}$, *and we find*

$$\begin{aligned} 0 &= \lambda_j h_j^{\nabla\rho_1\ldots\rho_n}(x^*)(\widetilde{\rho}(t^*) - t^*) \\ &\quad + \int_{t^*}^{\widetilde{\rho}(t^*)} \left(\lambda_j H_{1j}^1(x(\tau))x^{\widetilde{\nabla}^2}(\tau) + \lambda_j x^{\widetilde{\nabla}}(\tau)^T H_{2j}^1(x(\tau))x^{\widetilde{\nabla}}(\tau)\right) h_1(\widetilde{\rho}(t^*),\rho(\tau))\widetilde{\nabla}\tau, \end{aligned}$$

$j \in \{1,\ldots,m\}$. *Therefore*

$$\begin{aligned} 0 &= \lambda^T h^{\nabla\rho_1\ldots\rho_n}(x^*)(\widetilde{\rho}(t^*) - t^*) \\ &\quad + \int_{t^*}^{\widetilde{\rho}(t^*)} \left(\lambda^T H_1^1(x(\tau))x^{\widetilde{\nabla}^2}(\tau) + x^{\widetilde{\nabla}}(\tau)^T(\lambda^T H_2^1(x(\tau)))x^{\widetilde{\nabla}}(\tau)\right) h_1(\widetilde{\rho}(t^*),\rho(\tau))\widetilde{\nabla}\tau. \end{aligned} \tag{5.31}$$

By the Taylor formula, applying for g, we obtain

$$\begin{aligned} 0 &\geq g_j(\rho(x^*)) - g_j(x^*) \\ &= g_j^{\nabla\rho_1\ldots\rho_n}(x^*)(\widetilde{\rho}(t^*) - t^*) \\ &\quad + \int_{t^*}^{\widetilde{\rho}(t^*)} \left(G_{1j}^1(x(\tau))x^{\widetilde{\nabla}^2}(\tau) + x^{\widetilde{\nabla}}(\tau)^T G_{2j}^1(x(\tau))x^{\widetilde{\nabla}}(\tau)\right) h_1(\widetilde{\rho}(t^*),\widetilde{\rho}(\tau))\widetilde{\nabla}\tau, \end{aligned}$$

$j \in \{1,\ldots,p\}$. *We multiply by* θ_j, $j \in \{1,\ldots,p\}$, *and we find*

$$0 \geq \theta_j g_j^{\nabla\rho_1\ldots\rho_n}(x^*)(\widetilde{\rho}(t^*) - t^*)$$

$$+\int_{t^*}^{\widetilde{\rho}(t^*)}\left(\theta_j G_{1j}^1(x(\tau))x^{\widetilde{\nabla}^2}(\tau)+\theta_j x^{\widetilde{\nabla}}(\tau)^T G_{2j}^1(x(\tau))x^{\widetilde{\nabla}}(\tau)\right)h_1(\widetilde{\rho}(t^*),\rho(\tau))\widetilde{\nabla}\tau,$$

$j\in\{1,\ldots,p\}$. *Therefore*

$$\begin{aligned}0 &\geq \theta^T g^{\nabla\rho_1\ldots\rho_n}(x^*)(\widetilde{\rho}(t^*)-t^*)\\ &\quad+\int_{t^*}^{\widetilde{\rho}(t^*)}\left(\theta^T G_1^1(x(\tau))x^{\widetilde{\nabla}^2}(\tau)+x^{\widetilde{\nabla}}(\tau)^T(\theta^T G_2^1(x(\tau)))x^{\widetilde{\nabla}}(\tau)\right)h_1(\widetilde{\rho}(t^*),\rho(\tau))\widetilde{\nabla}\tau.\end{aligned}\tag{5.32}$$

Now, we apply the Taylor formula for f *and we find*

$$\begin{aligned}0 &> f(\rho(x^*))-f(x^*)\\ &= f^{\nabla\rho_1\ldots\rho_n}(x^*)(\widetilde{\rho}(t^*)-t^*)\\ &\quad+\int_{t^*}^{\widetilde{\rho}(t^*)}\left(F_1^1(x(\tau))x^{\widetilde{\nabla}^2}(\tau)+x^{\widetilde{\nabla}}(\tau)^T F_2^1(x(\tau))x^{\widetilde{\nabla}}(\tau)\right)h_1(\widetilde{\rho}(t^*),\rho(\tau))\widetilde{\nabla}\tau.\end{aligned}$$

We add the last inequality, (5.31) *and* (5.32), *and we arrive at*

$$\begin{aligned}0 &> \lambda^T h^{\nabla\rho_1\ldots\rho_n}(x^*)(\widetilde{\rho}(t^*)-t^*)\\ &\quad+\int_{t^*}^{\widetilde{\rho}(t^*)}\left(\lambda^T H_1^1(x(\tau))x^{\widetilde{\nabla}^2}(\tau)+x^{\widetilde{\nabla}}(\tau)^T(\lambda^T H_2^1(x(\tau)))x^{\widetilde{\nabla}}(\tau)\right)h_1(\widetilde{\rho}(t^*),\rho(\tau))\widetilde{\nabla}\tau\\ &\quad+\theta^T g^{\nabla\rho_1\ldots\rho_n}(x^*)(\widetilde{\rho}(t^*)-t^*)\\ &\quad+\int_{t^*}^{\widetilde{\rho}(t^*)}\left(\theta^T G_1^1(x(\tau))x^{\widetilde{\nabla}^2}(\tau)+x^{\widetilde{\nabla}}(\tau)^T(\theta^T G_2^1(x(\tau)))x^{\widetilde{\nabla}}(\tau)\right)h_1(\widetilde{\rho}(t^*),\rho(\tau))\widetilde{\nabla}\tau\\ &\quad+f^{\nabla\rho_1\ldots\rho_n}(x^*)(\widetilde{\rho}(t^*)-t^*)\\ &\quad+\int_{t^*}^{\widetilde{\rho}(t^*)}\left(F_1^1(x(\tau))x^{\widetilde{\nabla}^2}(\tau)+x^{\widetilde{\nabla}}(\tau)^T F_2(x(\tau))x^{\widetilde{\nabla}}(\tau)\right)h_1(\widetilde{\rho}(t^*),\rho(\tau))\widetilde{\nabla}\tau\\ &= \left(f^{\nabla\rho_1\ldots\rho_n}(x^*)+\lambda^T h^{\nabla\rho_1\ldots\rho_n}(x^*)+\theta^T g^{\nabla\rho_1\ldots\rho_n}(x^*)\right)(\widetilde{\rho}(t^*)-t^*)\\ &\quad+\int_{t^*}^{\widetilde{\rho}(t^*)}\Bigg((F_1^1(x(\tau))+\lambda^T H_1^1(x(\tau))+\theta^T G_1^1(x(\tau)))x^{\widetilde{\nabla}^2}(\tau)\end{aligned}$$

$$+x^{\widetilde{\nabla}}(\tau)^T\left(F_2^1(x(\tau))+\lambda^T H_2^1(x(\tau))+\theta^T G_2^1(x(\tau))\right)x^{\widetilde{\nabla}}(\tau)\Big)h_1(\widetilde{\rho}(t^*),\rho(\tau))\widetilde{\nabla}\tau$$

$$\geq \quad 0.$$

This is a contradiction.

4. *Let x^* be left-dense. Then there is a sequence $\{y^k\}_{k\in\mathbb{N}}$ such that $y^k \to x^*$, as $k\to\infty$, $y^k = x(t^k)$, $t^k < t^*$, $k\in\mathbb{N}$, and y^k, $k\in\mathbb{N}$, are feasible, and $y^k < x^*$, $k\in\mathbb{N}$, and*

$$f(y^k) < f(x^*), \quad k\in\mathbb{N}.$$

We have

$$\begin{aligned} h_j(y^k) &= 0, \\ h_j(x^*) &= 0, \quad j\in\{1,\ldots,m\}. \end{aligned}$$

Then

$$0 = h_j(y^k)-h_j(x^*), \quad j\in\{1,\ldots,m\}.$$

On the other hand, by the Taylor formula, we get

$$\begin{aligned} 0 &= h_j(y^k)-h_j(x^*) \\ &= h_j^{\nabla\rho_1\ldots\rho_n}(x^*)(t^k-t^*) \\ &\quad +\int_{t^*}^{t^k}\left(H_{1j}^1(x(\tau))x^{\widetilde{\nabla}^2}(\tau)+x^{\widetilde{\nabla}}(\tau)^T H_{2j}^1(x(\tau))x^{\widetilde{\nabla}}(\tau)\right)h_1(t^k,\rho(\tau))\widetilde{\nabla}\tau, \end{aligned}$$

$j\in\{1,\ldots,m\}$. We multiply by λ_j, $j\in\{1,\ldots,m\}$, and we find

$$\begin{aligned} 0 &= \lambda_j h_j^{\nabla\rho_1\ldots\rho_n}(x^*)(t^k-t^*) \\ &\quad +\int_{t^*}^{t^k}\left(\lambda_j H_{1j}^1(x(\tau))x^{\widetilde{\nabla}^2}(\tau)+\lambda_j x^{\widetilde{\nabla}}(\tau)^T H_{2j}^1(x(\tau))x^{\widetilde{\nabla}}(\tau)\right)h_1(t^k,\rho(\tau))\widetilde{\nabla}\tau, \end{aligned}$$

$j\in\{1,\ldots,m\}$. Therefore

$$\begin{aligned} 0 &= \lambda^T h^{\nabla\rho_1\ldots\rho_n}(x^*)(t^k-t^*) \\ &\quad +\int_{t^*}^{t^k}\left(\lambda^T H_1^1(x(\tau))x^{\widetilde{\nabla}^2}(\tau)+x^{\widetilde{\nabla}}(\tau)^T(\lambda^T H_2^1(x(\tau)))x^{\widetilde{\nabla}}(\tau)\right)h_1(t^k,\rho(\tau))\widetilde{\nabla}\tau. \end{aligned} \tag{5.33}$$

By the Taylor for the constraint g, we find

$$\begin{aligned} 0 &\geq g_j(y^k) - g_j(x^*) \\ &= g_j^{\nabla\rho_1\ldots\rho_n}(x^*)(t^k - t^*) \\ &\quad + \int_{t^*}^{t^k} \left(G_{1j}^1(x(\tau)) x^{\widetilde{\nabla}^2}(\tau) + x^{\widetilde{\nabla}}(\tau)^T G_{2j}^1(x(\tau)) x^{\widetilde{\nabla}}(\tau) \right) h_1(t^k, \rho(\tau)) \widetilde{\nabla}\tau, \end{aligned}$$

$j \in \{1, \ldots, p\}$. *We multiply by* θ_j, $j \in \{1, \ldots, p\}$, *and we find*

$$\begin{aligned} 0 &\geq \theta_j g_j^{\nabla\rho_1\ldots\rho_n}(x^*)(t^k - t^*) \\ &\quad + \int_{t^*}^{t^k} \left(\theta_j G_{1j}^1(x(\tau)) x^{\widetilde{\nabla}^2}(\tau) + \theta_j x^{\widetilde{\nabla}}(\tau)^T G_{2j}^1(x(\tau)) x^{\widetilde{\nabla}}(\tau) \right) h_1(t^k, \rho(\tau)) \widetilde{\nabla}\tau, \end{aligned}$$

$j \in \{1, \ldots, p\}$. *Therefore*

$$\begin{aligned} 0 &\geq \theta^T g^{\nabla\rho_1\ldots\rho_n}(x^*)(t^k - t^*) \\ &\quad + \int_{t^*}^{t^k} \left(\theta^T G_1^1(x(\tau)) x^{\widetilde{\nabla}^2}(\tau) + x^{\widetilde{\nabla}}(\tau)^T (\theta^T G_2^1(x(\tau))) x^{\widetilde{\nabla}}(\tau) \right) h_1(t^k, \rho(\tau)) \widetilde{\nabla}\tau. \end{aligned} \tag{5.34}$$

Now, we apply the Taylor formula for f *and we find*

$$\begin{aligned} 0 &> f(y^k) - f(x^*) \\ &= f^{\nabla\rho_1\ldots\rho_n}(x^*)(t^k - t^*) \\ &\quad + \int_{t^*}^{t^k} \left(F_1^1(x(\tau)) x^{\widetilde{\nabla}^2}(\tau) + x^{\widetilde{\nabla}}(\tau)^T F_2^1(x(\tau)) x^{\widetilde{\nabla}}(\tau) \right) h_1(t^k, \rho(\tau)) \widetilde{\nabla}\tau. \end{aligned}$$

We add the last inequality and (5.19) *and we arrive at*

$$\begin{aligned} 0 &> \lambda^T h^{\nabla\rho_1\ldots\rho_n}(x^*)(t^k - t^*) \\ &\quad + \int_{t^*}^{t^k} \left(\lambda^T H_1^1(x(\tau)) x^{\widetilde{\nabla}^2}(\tau) + x^{\widetilde{\nabla}}(\tau)^T (\lambda^T H_2(x(\tau))) x^{\widetilde{\nabla}}(\tau) \right) h_1(t^k, \rho(\tau)) \widetilde{\nabla}\tau \end{aligned}$$

$$
\begin{aligned}
& +\theta^T g^{\nabla\rho_1\ldots\rho_n}(x^*)(t^k-t^*) \\
& +\int_{t^*}^{t^k}\left(\theta^T G_1^1(x(\tau))x^{\widetilde{\nabla}^2}(\tau)+x^{\widetilde{\nabla}}(\tau)^T(\theta^T G_2^1(x(\tau)))x^{\widetilde{\nabla}}(\tau)\right)h_1(t^k,\rho(\tau))\widetilde{\nabla}\tau \\
& +f^{\nabla\rho_1\ldots\rho_n}(x^*)(t^k-t^*) \\
& +\int_{t^*}^{t^k}\left(F_1^1(x(\tau))x^{\widetilde{\nabla}^2}(\tau)+x^{\widetilde{\nabla}}(\tau)^T F_2^1(x(\tau))x^{\widetilde{\nabla}}(\tau)\right)h_1(t^k,\rho(\tau))\widetilde{\nabla}\tau \\
= & \left(f^{\nabla\rho_1\ldots\rho_n}(x^*)+\lambda^T h^{\nabla\rho_1\ldots\rho_n}(x^*)+\theta^T g^{\nabla\rho_1\ldots\rho_n}(x^*)\right)(t^k-t^*) \\
& +\int_{t^*}^{t^k}\Bigg(\left(F_1^1(x(\tau))+\lambda^T H_1^1(x(\tau))+\theta^T G_1^1(x(\tau))\right)x^{\widetilde{\nabla}^2}(\tau) \\
& +x^{\widetilde{\nabla}}(\tau)^T\left(F_2^1(x(\tau))+\lambda^T H_2^1(x(\tau))+\theta^T G_2^1(x(\tau))\right)x^{\widetilde{\nabla}}(\tau)\Bigg)h_1(t^k,\rho(\tau))\widetilde{\nabla}\tau \\
\geq & \ 0.
\end{aligned}
$$

This is a contradiction. This completes the proof.

As above, one can prove the following results regarding the local maximum points of the objective function.

Theorem 47 *(Second Order Sufficiency Conditions) Suppose that there is a point x^* such that*

$$h(x^*)=0$$

and a $\lambda\in\mathbb{R}^m$ such that

$$
\begin{aligned}
\theta^T g(x^*) &= 0, \\
f^{\Delta\sigma_1\ldots\sigma_n}(x^*)+\lambda^T h^{\Delta\sigma_1\ldots\sigma_n}(x^*)+\theta^T g^{\Delta\sigma_1\ldots\sigma_n}(x^*) &\leq 0, \\
f^{\nabla\rho_1\ldots\rho_n}(x^*)+\lambda^T h^{\nabla\rho_1\ldots\rho_n}(x^*)+\theta^T g^{\nabla\rho_1\ldots\rho_n}(x^*) &\geq 0, \\
F_1(x^*)+\lambda^T H_1(x^*)+\theta^T G_1(x^*) &\leq 0,
\end{aligned}
$$

$$F_1^1(x^*) + \lambda^T H_1^1(x^*) + \theta^T G_1^1(x^*) \quad \geq \quad 0.$$

Suppose also, that

$$L(x^*) = F_2(x^*) + \lambda^T H_2(x^*) + \theta^T G_2(x^*)$$

and

$$L^1(x^*) = F_2^1(x^*) + \lambda^T H_2^1(x^*) + \theta^T G_2^1(x^*)$$

are negative definite on

$$M = \{y \in \Lambda^m : h^{\Delta\sigma_1 \ldots \sigma_n}(x^*)y = 0, \quad g^{\Delta\sigma_1 \ldots \sigma_n}(x^*)y = 0\}$$

and

$$M^1 = \{y \in \Lambda^m : h^{\nabla\rho_1 \ldots \rho_n}(x^*)y = 0, \quad g^{\nabla\rho_1 \ldots \rho_n}(x^*)y = 0\},$$

respectively, that is

$$y^T L(x^*)y < 0$$

and

$$z^T L^1(x^*)z < 0,$$

respectively, for any $y \in M$, $y \neq 0$ *and* $z \in M^1$, $z \neq 0$. *Then* x^* *is a local maximum point for the objective function* f *subject to*

$$h(x) = 0, \quad g(x) \leq 0.$$

Index